Aroma

Smell is a *social* phenomenon, invested with particular meanings and values by different cultures. Odours form the building blocks of cosmologies, class hierarchies and political orders; they can enforce social structures or transgress them, unite people or divide them, empower or disempower.

But smell is repressed in the modern West, and its social history ignored. *Aroma* breaks the 'olfactory silence' of modernity by offering the first comprehensive exploration of the cultural role of odours in Western history – from antiquity to the present – and in a wide variety of non-Western societies. Its topics range from the medieval concept of the 'odour of sanctity' to the aroma-therapies of South America, and from olfactory stereotypes of gender and ethnicity in the modern West to the role of smell in postmodernity.

Aroma will make essential reading for students of cultural studies, history, anthropology and sociology. Its engaging style and heady subject matter are sure to fascinate anyone who likes to nose around in the hothouse of culture.

Constance Classen is the author of *Inca Cosmology and the Human Body* and *Worlds of Sense*; **David Howes** is the editor of *The Varieties of Sensory Experience*; and **Anthony Synnott** is the author of *The Body Social*.

Aroma

The cultural history of smell

Constance Classen, David Howes and Anthony Synnott

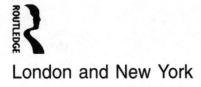

London and New York

First published 1994
by Routledge
11 New Fetter Lane, London EC4P 4EE

Simultaneously published in the USA and Canada
by Routledge
29 West 35th Street, New York, NY 10001

Reprinted in 1997

Phototypeset in Times by Intype, London

Printed and bound in Great Britain by
Mackays of Chatham PLC, Chatham, Kent

British Library Cataloguing in Publication Data
A catalogue record for this book is available from the
British Library.

Library of Congress Cataloging in Publication Data
Classen, Constance,
    Aroma: the cultural history of smell/Constance Classen,
    David Howes, and Anthony Synnott.
    p. cm.
    Includes bibliographical references and index.
    1. Smell – History. 2. Odors – History. 3. Social history.
    I. Howes, David, 1957– . II. Synnott, Anthony, 1940–
    III. Title
    GT2847.C53   1994
    391´.63 – dc20
                                                    94–5592
                                                    CIP

ISBN 0-415-11472-1 (hbk)
ISBN 0-415-11473-X (pbk)

Contents

Tables

Acknowledgements

This book is the result of several years of study of the social role of smell in the West and across cultures. The idea to write a book on odour originated when the three of us participated in a research project on 'The Varieties of Sensory Experience' based at Concordia University, Montreal from 1988 to 1991. From that period on we have been gathering material on smell in culture – more material, indeed, than we have been able to use here – and working out how to order and interpret our findings. Our research was funded in part by the Olfactory Research Fund, New York, and we would particularly like to thank Annette Green of the Olfactory Research Fund for her support. The opinions expressed in this book are, of course, entirely our own.

Constance Classen would like to thank Gregory Baum and Lionel Sanders for their advice and encouragement. A fellowship from the Social Sciences and Humanities Research Council of Canada enabled her to research and write her contribution to this book. She is indebted to Lawrence Sullivan for making it possible for her to carry out her research as a fellow of the Center for the Study of World Religions, Harvard University.

David Howes would like to thank Michael Lambek, Alain Corbin, Marguerite Dupire, Rob Shields, Gail Guthrie Valaskakis and the students in his psychological anthropology classes at Concordia University for the encouragement and highly constructive criticisms they have provided.

Anthony Synnott would like to thank the faculty and students in the Department of Sociology and Anthropology at Concordia, who participated so willingly in the Concordia smell survey, as well as Nicolette Starkie, who worked so hard on the survey, and

Carole Robertson for typing his share of the manuscript; special thanks to Hélène Tobin.

We are all indebted to George Classen for his assistance in preparing the manuscript for publication, and to Chris Rojek and our editors at Routledge for their support and guidance.

Chapters 1 and 2 were researched and written by Constance Classen, while the other chapters represent more of a collective effort. Chapter 6 is taken from C. Classen and D. Howes, 'L'arôme de la marchandise', *Anthropologie et Sociétés*, 1994, vol. 18, no. 3.

Introduction
The meaning and power of smell

Smell is powerful. Odours affect us on a physical, psychological and social level. For the most part, however, we breathe in the aromas which surround us without being consciously aware of their importance to us. It is only when our faculty of smell is impaired for some reason that we begin to realize the essential role olfaction plays in our sense of well-being. One man who lost his sense of smell due to a head injury expressed this realization as follows:

> when I lost [my sense of smell] – it was like being struck blind. Life lost a good deal of its savour – one doesn't realize how much 'savour' is smell. You *smell* people, you *smell* books, you *smell* the city, you *smell* the spring – maybe not consciously, but as a rich unconscious background to everything else. My whole world was suddenly radically poorer.[1]

A survey conducted by Anthony Synnott at Montreal's Concordia University asked 270 students and professors to comment on the role of smell in their lives.[2] The question, 'What are your favourite smells?' elicited a wide range of responses, from the expected – 'the smell of babies', 'freshly mown lawn', 'roses', 'home-made bread' – to the unexpected: 'the odours of the Montreal Forum and the Olympic Stadium', 'body perspiration', 'dogs', 'gasoline'. The question, 'Which smells do you dislike?' evoked a similar variety of responses: 'smelly men on the bus', 'pig farms and chicken coops', 'cigarette smoke', 'hospitals', 'raw meat'. Interestingly, while for many people commercial perfumes had fond associations, many listed them among the odours they disliked. Some stressed the physical discomfort perfumes gave them: 'instant headache and nausea', said one respondent of her reaction; 'perfumes make me sneeze', said another. Others complained that perfumes obscured natural odours and desensitized the senses in general.[3]

Smell can evoke strong emotional responses. A scent associated with a good experience can bring a rush of joy. A foul odour or one associated with a bad memory may make us grimace with disgust. Respondents to the survey noted that many of their olfactory likes and dislikes were based on emotional associations. Such associations can be powerful enough to make odours that would generally be labelled unpleasant agreeable, and those that would generally be considered fragrant disagreeable for particular individuals. The smell of gasoline, for example, usually thought to be unpleasant, was enjoyed by one respondent because '[it] reminds me of all the places I can go and have been in my car, i.e. freedom'. The smell of sports stadiums was a preferred scent of another because he associated it with his favourite sports. Likewise some seemingly innocuous or pleasant scents, such as carrots, cantaloupe and flowers, were strongly disliked by certain respondents because of the bad experiences associated with them:

> When my father passed away two years ago, we put a certain kind of flower in front of his picture. That same kind of smell reminds me of the sadness, the helplessness, worst of all my mother's crying.

The perception of smell, thus, consists not only of the sensation of the odours themselves, but of the experiences and emotions associated with them.[4]

Odours are essential cues in social bonding. One respondent to the smell survey noted, 'I think there is no true *emotional* bonding without touching and smelling, burying one's nose into a loved one.' In fact, infants recognize the odours of their mothers soon after birth and adults can identify their children or spouses by scent. In one well-known test, women and men were able to distinguish t-shirts worn by their marriage partners – from among dozens of others – by smell alone.[5] Most of these subjects would probably never have given much thought to odour as a cue for identifying family members before being involved in the test, but as the experiment revealed, even when not consciously considered, smells *register*.

In spite of its importance to our emotional and sensory lives, smell is probably the most undervalued sense in the modern

West.[6] The reason often given for the low regard in which smell is held is that, in comparison with its importance among animals, the human sense of smell is feeble and atrophied. While it is true that the olfactory powers of humans are nothing like as fine as those possessed by certain animals, they are still remarkably acute. Our noses are able to recognize thousands of smells, and to perceive odours which are present only in infinitesimally small quantities.

Smell, however, is a highly elusive phenomenon. Odours, unlike colours, for instance, cannot be named – at least not in European languages. 'It smells like . . .', we have to say when describing an odour, groping to express our olfactory experience by means of metaphors. Nor can odours be recorded: there is no effective way of either capturing scents or storing them over time. In the realm of olfaction, we must make do with descriptions and recollections.

Most of the research on smell undertaken to date has been of a physical scientific nature. Significant advances have been made in the understanding of the biological and chemical nature of olfaction, but many fundamental questions have yet to be answered: is smell one sense or two – one responding to odours proper and the other registering odourless pheromones (air-borne chemicals)? Is the nose the only part of the body affected by odours? How can smells be measured objectively? There is also a body of research in the psychology of smell. Various experiments have been done in an attempt to find out the effects of odours on the performance of tasks, on mood, on dieting, and so on.[7]

Smell is not simply a biological and psychological phenomenon, though. Smell is *cultural*, hence a social and historical phenomenon.[8] Odours are invested with cultural values and employed by societies as a means of and model for defining and interacting with the world. The intimate, emotionally charged nature of the olfactory experience ensures that such value-coded odours are interiorized by the members of society in a deeply personal way. The study of the cultural history of smell is, therefore, in a very real sense, an investigation into the *essence* of human culture.

The devaluation of smell in the contemporary West is directly linked to the revaluation of the senses which took place during the eighteenth and nineteenth centuries. The philosophers and scientists of that period decided that, while sight was the pre-eminent sense of reason and civilization, smell was the sense of madness and savagery. In the course of human evolution, it was argued by Darwin, Freud and others, the sense of smell had been left behind and that of sight had taken priority.[9] Modern humans who emphasized the importance of smell were therefore judged to be either insufficiently evolved savages, degenerate proletariat, or else aberrations: perverts, lunatics or idiots.

This powerful denigration of smell by Europe's intellectual elite has had a lasting effect on the status of olfaction. Smell has been 'silenced' in modernity. Even on those rare occasions when it *is* the subject of popular discourse – for example, in certain contemporary works of fiction – it tends to be presented in terms of its stereotypical association with moral and mental degeneracy.

Patrick Süskind's enormously popular book *Perfume* is a case in point. The keen-scented protagonist of the book, Jean-Baptiste Grenouille, is both 'idiot' and 'pervert' – as well as an offspring of the 'degenerate' lower class. Grenouille exercises his abnormal passion for scent by murdering maidens in order to sniff up their sweet fragrance. In the end, through his de-scenting of maidens, Grenouille is able to invest himself with an odour so attractive that he is torn to pieces and eaten by a frenzied crowd.[10]

If *Perfume* makes for a 'good read', it is not only because of its unusual topic and engrossing story line, but also (and perhaps more fundamentally) because of its confirmation of the validity of many of our most cherished olfactory stereotypes – the maniac sniffing out his prey; the fragrant, hapless maiden; the dangerous savagery inherent in the sense of smell.

Why this cultural repression and denigration of smell? Generally speaking, those elements which are systematically suppressed by a culture are so regulated not only because they are considered inferior, but also because they are considered threatening to the social order. In what ways, one wonders, could a heightened olfactory consciousness be dangerous to the established social order in the West?

For one thing, in the premodern West, odours were thought of as intrinsic 'essences', revelatory of inner truth. Through smell, therefore, one interacted with *interiors*, rather than with surfaces, as one did through sight. Furthermore, odours cannot be readily contained, they escape and cross boundaries, blending different entities into olfactory wholes. Such a sensory model can be seen to be opposed to our modern, linear worldview, with its emphasis on privacy, discrete divisions, and superficial interactions.

This is not to suggest that an olfactory-minded society would be an egalitarian utopia with all members harmoniously combining into a cultural perfume. As we shall see, olfactory codes can and often do serve to divide and oppress human beings, rather than unite them. The suggestion is rather that smell has been marginalized because it is felt to threaten the abstract and impersonal regime of modernity by virtue of its radical interiority, its boundary-transgressing propensities and its emotional potency. Contemporary society demands that we distance ourselves from the emotions, that social structures and divisions be seen to be objective or rational and not emotional, and that personal boundaries be respected. Thus, while olfactory codes continue to be allowed to reinforce social hierarchies at a semi- or subconscious level, sight, as the most detached sense (by Western standards), provides *the* model for modern bureaucratic society.[11]

Academic studies of smell have tended to suffer from the same cultural disadvantages as smell itself. While the high status of sight in the West makes it possible for studies of vision and visuality, even when they are critical, to be taken seriously, any attempt to examine smell runs the risk of being brushed off as frivolous and irrelevant. None the less, the role of odour in culture is such a profound and fascinating subject that a number of scholars in different fields – including history, sociology and anthropology – have sought to explore it in their work. The present book brings together some of the contributions of these 'pioneers' in the cultural study of smell, as well as integrating relevant data from a wide range of other sources, with the purpose of providing a historical and cross-cultural account of beliefs and practices concerning smell. *Aroma*, indeed, offers the first comprehensive exploration of the *cultural* role of odours in different periods of Western history up to and including the present, and in a wide range of non-Western societies.

The first part of *Aroma*, 'In Search of Lost Scents', presents an 'archaeology' of smell in an attempt to recover – under the many layers of contemporary Western visualism – the olfactory world of the premodern West. Chapter 1, 'The Aromas of Antiquity', deals with the role of odour in classical times. Paul Fauré writes in *Parfums et aromates de l'antiquité* that our sense of smell is so underdeveloped in the modern West that we can no more appreciate the importance of odour in the ancient world than the blind can describe a colourful scene.[12] The ancients made rich use indeed of aromatics, from the spiced pageantry of royal parades through the perfumed banquets of the wealthy to the incensed temples of the gods.

Just as important as the actual use of odour in the ancient world, however, were its metaphorical and literary uses. The range of classical olfactory expressions – in the form of quips, paeans, and condemnations – comes through vividly, even after so many centuries, in the writings of contemporary playwrights and poets. Consider, for example, the lyrical beauty of the following olfactory evocation of a kiss by the Roman epigrammatist Martial:

> Breath of balm from phials of yesterday, of the last effluence that falls from a curving jet of saffron; perfume of apples ripening in their winter chest, of fields lavish with the leafage of spring . . . [13]

These lovingly crafted lines on scent are an indication of the intimate meaning odours had for the ancients.

The second chapter, 'Following the Scent: From the Middle Ages to Modernity', picks up the scent trail of the West after the fall of the Roman Empire. The combined influences of Christian asceticism and barbarian austerity led to a decline in the use of perfumes after the collapse of Rome. With the Crusades, however, the peoples of the West were once again brought into contact on a large scale with the spices and perfumes of the East which had so entranced the Greeks and Romans. Aromatics were an essential part of the good life of medieval to Enlightenment Europe. So much so that court etiquette in seventeenth-century Versailles, for example, demanded that a different scent be worn each day of the week.

At the same time, fundamental spiritual and curative powers were attributed to scent by Christendom. These special powers of smell can be seen in such contemporary concepts as 'the odour of sanctity' and in the role played by aromatics during periods of plague, when the battle against disease must often have appeared to be a war waged between fragrant and foul scents. By the nineteenth century, however, following what Alain Corbin has called the 'olfactory revolution',[14] fragrance had moved out of the realms of religion and medicine into those of sentiment and sensuality. This move is brought out in the works of many writers of that period, such as Baudelaire and, later, Proust, who used olfactory symbolism in their writings to create an evocative atmosphere. The final section of Chapter 2 examines the attitudes of nineteenth-century thinkers – from Darwin to Freud – towards odour and explores their influence on the olfactory norms of the modern West.

Part II, entitled 'Explorations in Olfactory Difference', compares the role of smell in various non-Western cultures. It opens with 'Universes of Odour'. This chapter deals with how smell is used to structure and classify different aspects of the world, from time and space to gender and selfhood. Examples are drawn from the 'osmologies', or olfactory classification systems, of cultures ranging from the Bororo of Brazil to the Dassanetch of Ethiopia. The chapter begins with an account of the 'calendar of scents' used by the aboriginal inhabitants of the Andaman Islands to reckon time, and moves on to examine the 'smellscapes' of various peoples of the rainforest. Other topics considered include the smell vocabularies of non-Western cultures and the use of olfactory codes as models for social organization. The chapter concludes with a discussion of how smell symbolism is linked to other sensory symbolic systems in certain cultures.

Aromatics are employed across cultures for a variety of purposes, including seduction, healing, hunting and communication with the spirits. Chapter 4 offers a comparison and analysis of some of the diverse 'rites of smell' which have been elaborated around these activities. Among the Umeda of New Guinea, for example, a hunter sleeps with a bundle of herbs tucked under his pillow, the aroma of which is supposed to inspire dreams of the chase. The next day he has only to act out his scent-inspired dream to enjoy a successful hunt. The Warao of Venezuela, who have developed a complex system of aromatherapy, enlist powerful herbal scents to combat the evil odours of disease. The Amazonian Desana use scents, along with other sensory stimuli, to help direct hallucinogenic visions. These and other ritual uses of scent translate the olfactory classification systems discussed in the previous chapter into practice.

The final part of the book is called 'Odour, Power and Society' and deals with the olfactory traits of the modern West. Through having developed an understanding of the social history of smell in the West and an appreciation of its cultural elaboration in non-

Western societies in the two previous parts, we are better able to penetrate the olfactory symbol systems of the contemporary West in this concluding section.

Chapter 5 is concerned with documenting 'the politics of smell'. The interrelationship between the olfactory and the political was highlighted by George Orwell in the early twentieth century when he proclaimed that the 'real secret' of class distinctions in the West could be summed up in 'four frightful words ... *The lower classes smell* ... No feeling of like or dislike is quite so fundamental as a *physical* feeling.' Orwell went on to assert that race hatred, religious hatred, differences of education, of temperament, can all be overcome, 'but physical repulsion cannot', whence the persistence of class distinctions.[15]

Orwell's point is striking, but while the feeling of physical repulsion to which he alludes appears fundamental, it is important to understand that its basis remains social rather than physical, since class divisions are given in society, not in nature. As olfactory preferences and aversions tend to take root deep in the human psyche, evoking or manipulating odour values is a common and effective means of generating and maintaining social hierarchies. This may explain why smell is enlisted not only to create and enforce class boundaries, but also ethnic and gender boundaries. Such olfactory social codes often pass unnoticed by us, for they tend to function below the surface of conscious thought. Our study of the politics of smell in modernity brings the interrelations of odour, power and society very much to the fore, however, by examining such topics as the 'scent-typing' of women, the olfactory symbolism of the Nazi concentration camp and the regulation of the odours of public space.

Chapter 6, 'The Aroma of the Commodity: The Commercialization of Smell', explores the production and regulation of odours in today's consumer culture. Olfactory management takes place on numerous levels: the body, the home, the workplace and the marketplace. At the level of the body, for instance, deodorants suppress unwanted odours while perfumes and colognes allow for the creation of an ideal olfactory image. At the level of the workplace, the concern is with how to develop an attractive olfactory atmosphere that will stimulate and refresh workers, as opposed to the stale air that is usually found in the enclosed modern office building. In the marketplace, businesses are increasingly concerned not only with new ways of marketing perfumes, such as home fragrance products and aromatherapy, but with the addition of synthetic fragrances to a variety of products, from processed foods to house paints. The chapter ends with a discussion of how odour, as it is increasingly simulated by fragrance engineers and commercialized by marketers, is passing from the realm of modernity to that of postmodernity.

When considered as a whole, the three parts of *Aroma* offer intriguing juxtapositions of olfactory beliefs and practices from different cultures and eras. In classical Greek cosmology, for example, an odour of spices was associated with the sun. The present-day Desana of Colombia attribute a honey-like sweetness to the sun. Among the Batek Negrito of Malaysia, however, the sun is thought to emit pathogenic foul odours. In modern Western cosmology, of course, the sun is basically a visual entity, with no olfactory identity.

At times such different beliefs and customs are seen to overlap as cultures interact with each other. Thus, while in Chapters 3 and 4 we learn of the traditional olfactory concepts and rituals of peoples of the Amazon, in Chapter 6 we read of Amazonian 'Avon ladies' who tour the isolated towns and villages of the Amazon trying to sell or barter such popular Avon products as 'Crystal Splash' cologne and 'Bart Simpson' deodorant. It is important to keep in mind, indeed, that the olfactory ethos of the modern West is by no means confined to the 'developed' countries of the First World, but is carrying its message across cultures on the wagon of consumer capitalism.

As an essay on the history, anthropology and sociology of odour, *Aroma* is necessarily restricted in the amount of space that can be devoted to any one topic. Our objective has been to be comprehensive rather than exhaustive. We hope that the present work will stimulate further research into the cultural construction of smell and, indeed, of all of the senses.

It might be argued that by focusing on smell to the exclusion of the other senses we have been guilty of sensory bias, and that the role of smell in culture can only be understood within a multisensory context. However, historians, anthropologists and sociologists have long excluded odour from their accounts and concentrated on the visual and the auditory, without being accused of any sensory biases. The argument must, therefore, be turned around. Our singling out of scent for attention serves to redress this long-standing imbalance, for we in the West tend to be so 'odour-blind' that unless smell is placed right under our noses, so to speak, it usually gets lost in the shuffle. By demonstrating the importance of odour and olfactory codes in both Western and non-Western societies, we wish to bring smell out of the Western scholarly and cultural unconscious into the open air of social and intellectual discourse. It is only when a form of sensory equilibrium has been recovered, that we may begin to understand how the senses interact with each other as models of perception and paradigms of culture.

Part I

In search of lost scents

Chapter 1

The aromas of antiquity

'The pleasure of perfume', wrote Pliny in the first century, '[is] among the most elegant and also most honourable enjoyments in life.'[1] The inhabitants of the ancient world, indeed, enjoyed sweet scents with an intensity which we moderns, for all the money we spend on perfumes, can scarcely imagine. People of antiquity used scent not only for purposes of personal attraction, but also as an important ingredient for everything from dinner parties through sporting events and parades to funerals. In our own age, by contrast, the notion of a perfumed dinner party or parade is so alien as to seem absurd.

Concomitantly, many of the foul smells which infused the lives of the inhabitants of earlier periods in Western history have been eliminated from our modern First World consciousness. In effect, therefore, an olfactory gulf lies between our own deodorized modern life and the richly scented lives of our forbears.[2] In what follows we will explore this lost world of scents in order to try and recapture, if only on paper, the essence of those earlier, more redolent, times.

The period focused on in this chapter is the first century AD. References are also made to selected works from earlier and later periods of antiquity, however, in order to indicate the continuity of certain beliefs and practices.

ATTAR OF ROSES, CINNAMON AND MYRRH

The Graces are described in classical poetry as wearing garments scented with

crocus, hyacinth, and blooming violet

and the sweet petals of the peerless rose
so fragrant, so divine.[3]

Such floral scents were, in fact, associated with a state of grace
in antiquity, evoking a sense of distilled youth and beauty: sweet,
fresh and evocative. Garlands and floral crowns were thought to
make fitting offerings for the gods, and to bestow on their wearers
an essence of divinity when worn by mortals.[4]

Interested as they were in floral scents, the ancients were particularly attentive to the olfactory nuances of the garden. Pliny
writes in his *Natural History*, for example,

> The smell of some plants is sweeter at a distance, becoming
> fainter as the distance is lessened; for instance, that of the
> violet. A freshly gathered rose smells at a distance, but a faded
> rose when nearer. All perfume however is stronger in spring,
> and in the morning; as the day draws near to noon it grows
> weaker. Young plants also have less perfume than old ones;
> the strongest perfume however of all plants is given out in
> middle age.[5]

Pliny also informs us that 'weather too makes a difference for in
certain years the rose grows with less perfume, and furthermore
all roses have more perfume on dry soils than on moist'.[6]

Not only the aromas of flowers, but also the odour of the earth
was appreciated by the ancients:

> It is certainly the case that a soil which has a taste of perfumes
> will be the best soil ... The earth [after a shower] sends out
> that divine breath of hers, of quite incomparable sweetness,
> which she has conceived from the sun. This is the odour which
> ought to be emitted when the earth is turned up, and the scent
> of the soil will be the best criterion of its quality.[7]

While many of the perfumes appreciated in Greece and Rome
could be found in the garden, others had to be imported from
Arabia. These included spices such as cinnamon and cassia, and
aromatic resins, such as myrrh and frankincense. The sale of
these aromatic products made the fortune of many an Arabian
merchant, as endless caravans carted loads of olfactory wealth
through the dusty deserts of Arabia *en route* to the markets of
Greece and Rome. *Arabia felix*, happy Arabia, was the name the

Romans wistfully gave to the country that produced such a fragrant bounty.[8]

Most of our information about the role of scent in the ancient world comes from the writers of Greece and Rome. The cultures of the ancient Middle East, however, had olfactory traditions that were in many respects more developed than those of their Western neighbours. The Greeks and Romans sometimes deprecated the extensive use of perfumes by the Egyptians, Persians and others as sensualist foppery. More than they deprecated, however, they admired and imitated. Just as aromatics travelled to Greece and Rome from the East, there is no doubt that many aromatic customs also came via the same route.

With their home-grown and their imported aromatics the ancients created gloriously heady blends of perfumes.[9] *Susinum*, made of lilies, oil of behen nut, sweet flag, honey, cinnamon, saffron and myrrh. *Megalium*, the great creation of the Roman perfumer Megallus, was made of balsam, rush, reed, behen nut oil, cassia and resin. The elaborate 'royal perfume' was composed of over twenty ingredients, including wild grape, spikenard, lotus, cinnamon, myrrh, gladiolus and marjoram. The most famous of Egyptian perfumes, *Kyphi*, was a blend of sixteen ingredients. According to the Greek historian Plutarch, this perfume had the power to relieve anxiety, brighten dreams, and heal the soul.[10] Kyphi was also a sacred incense, offered up by the inhabitants of the city of Heliopolis to the sun-god Re as he set in the sky every evening.[11]

As in our own fashion-conscious age, however, trends in perfumery came and went in the ancient world. 'The first thing proper to know about [perfumes]', writes Pliny, 'is that their importance changes.' Thus: 'The iris perfume of Corinth was extremely popular for a long time, but afterwards that of Cyzicus ... vine-flower scent made in Cyprus was preferred, but afterwards that from Adramytteum, and scent of marjoram made in Cos, but afterwards quince-blossom unguent.'[12] Our olfactory appetite is whetted by the thought of what the iris perfume of Corinth that was so 'extremely popular' was like, or the quince-blossom unguent from Cos or the vine-flower scent of Cyprus.

It is not only the ingredients of ancient perfumes that sound exotic to us now, but also the ways in which they were prepared. Scents were available in a variety of forms: as toilet waters or oils, as dry powders, in thick unguents, or as incense. Whereas when we think of perfumes today, we inevitably imagine them as liquids, an inhabitant of the ancient world would be just as likely to enjoy perfume in the form of a thick ointment, to be smeared liberally on the body, or a fragrant smoke, infusing the air with its odour. Our own English word 'perfume', in fact, literally means 'to smoke through', indicating the importance this method of imparting fragrance had for our ancestors.

As in our day, the well-to-do of antiquity bought their scents from perfumers. In one Greek play, for example, a perfumer named Peron is mentioned: 'I left the man in Peron's shop just now dealing for ointments. When he has agreed he'll bring your cinnamon and spikenard essence.'[13] Perfumers stored their wares in lead or alabaster vases to prevent their odours from evaporating. These vessels were kept in shady upper rooms of the shop where they would be shielded from the damaging heat of the sun. Clients shopping for a scent would have their wrists anointed with different oils by the perfumer, for, then as now, it was held that perfumes were sweetest when the scent came from the wrist. Perfumers employed other tricks of the trade to sell their wares as well. The early Greek botanist Theophrastus tells us, for instance, that the scent of roses is so powerful that it will overwhelm most other perfumes. Perfumers wishing their clients to buy attar of roses, therefore, would scent them with it first, after which all other perfumes they tried would seem relatively odourless.[14]

The royalty of antiquity had perfumers attached to their courts, not only to prepare perfumes for their own persons, but also for state feasts and entertainments. The amounts of perfumes and fragrant flowers used on such occasions could be enormous. Thus Darius III, King of Persia, for example, had in his retinue fourteen perfumers and forty-six garland makers.[15]

Perfumes were worn by both men and women on their hair, their breast and sometimes on legs and feet. The account in the New Testament of Jesus having his feet perfumed with expensive ointment provides a well-known example of this last custom. The true perfume lovers of antiquity were not content to anoint themselves with simply one scent, however, but would use different perfumes for different parts of the body. Antiphanes, in reference to this fashion, writes of a wealthy Greek who

... steeps his feet
And legs in rich Egyptian unguents;
His jaws and breasts he rubs with thick palm oil,
And both his arms with extract sweet of mint;
His eyebrows and his hair with marjoram,
His knees and neck with essence of ground thyme.[16]

Nothing less than a complete olfactory wardrobe! Such a discriminating use of perfumes indicates that the ancients were not simply
content to douse themselves with one strong scent or another,
but had a highly developed sense of olfactory aesthetics.

SCENTS OF THE CITY

Writing in the fifth century BC, Sophocles describes the city of
Thebes as being 'heavy with a mingled burden of sounds and
smells, of groans and hymns and incense.'[17] The cities and towns
of the ancient world did indeed offer a rich mélange of olfactory
and other sensations. Walking through the streets of Nero's Rome
in the first century AD, one would encounter the stench of refuse
rotting by the wayside, the piercing fragrance of burning myrrh
emanating from temples, the heavy aroma of food being cooked
by street vendors, the sweet, seductive scents of flowering gardens,
the malodour of rotting fish at a fishstand, the sharp smell of
urine from a public latrine and perhaps the incense trail of a
passing procession honouring a god or hero.[18]

Certain parts of the city had the characteristic scent of the
activities that were carried out there. A character in one of
Plautus's plays speaks of looking for someone in 'the squares,
gymnasiums, the barbers' shops, the mart, the shambles, and the
wrestling school, the forum, and the street where doctors dwell,
the perfume-sellers, all the sacred shrines.'[19] All these places
would have had their own distinctive odours throughout the
classical era: the gymnasium would smell of oil and sweat;
the markets of the produce sold there; the barber and perfume
shops of fragrant ointments; the shrines of incense and burnt
offerings.

Some places were particularly well known for their foul odours,
for example the tanneries, where nauseating-smelling hides were
made into leather, and the laundries, where fullers – washers and
dyers of clothes – used large quantities of urine as a cleansing

agent. Some places, in turn, were characterized by their fragrance, for example temples. Indeed, fragrance was such an important element of temples that not only were they heavily scented within, but perfume was occasionally mixed right into the mortar. Pliny, for example, writes that

> At Elis there is a temple of Minerva in which, it is said, Panaenus, the brother of Pheidias, applied plaster that had been worked with milk and saffron. The result is that even today, if one wets one's thumb with saliva and rubs it on the plaster, the latter still gives off the smell and taste of saffron.[20]

These different local odours created the effect of an olfactory map, enabling the inhabitants of the city to conceptualize their environment by way of smell.

When the citizens of Rome wished to cleanse themselves of the odours and grime of the city, which they customarily did once a day, they retired to the public baths. There, they could work up a sweat in the *sudatorium*, have a warm bath in the *tepidarium*, and then cool off with a swim in the cold water of the *frigidarium*. When finished, the bathers entered the *unctuarium*, anointing room, where those who could afford it were massaged and anointed with perfumes by slaves. After passing through the various chambers of the baths, the refreshed Roman citizen could go out into the city again in good odour, clean and sweet-scented.

HOUSEHOLD SCENTS

In some ways the homes of antiquity, with food cooking in the kitchen, incense burning on the household altar, garbage rotting in the waste pile, were olfactory models in miniature of the city. The homes and possessions of the well-to-do, however, were often perfumed with the same care as their own persons were. The walls of rooms would be daubed with perfumed unguent and the mosaic floors sprinkled with fragrant water and strewn with flowers. When the weather was cold, fires of scented wood would keep the homes of the wealthy both warm and perfumed.

Likewise, cushions might be filled with dried herbs, and powdered scents placed between the bedclothes to render sweet the hours of repose. Clothes were incensed with perfume and stored in chests of fragrant wood together with aromatics. Bath and pool water might be perfumed as well. The first-century Roman epigrammatist Martial, describing the comforts of one of his well-off contemporaries writes 'your sea-bath is pale with the tinge of your perfumes'.[21] Pliny, in turn, comments that

> we have heard that somebody of private station gave orders for the walls of his bathroom to be sprinkled with scent, and that the Emperor Caligula had the bathtubs scented, and so also later did one of the slaves of Nero – so that this must not be considered a privilege of princes![22]

So widespread was the use of perfume in the Roman household, that even domestic animals such as dogs and horses might find themselves perfumed with their owner's favourite scent, as in the following prescription for a favoured pet: 'Strew, then, soft carpets underneath the dog ... and with Megallian oils anoint his feet.'[23]

Olfactory practices of this sort, of course, could only be indulged in by the well-to-do. At the other extreme, the unkempt home of a poor family, living in crowded, dirty conditions, might smell very differently. A sense of what this smell would be like can be gathered from Martial's description of a poor family's household goods:

> a cracked chamberpot was making water through its broken side ... that there were salted gudgeons, too, or worthless sprats, the obscene stench of a jug confessed – such a stench as a whiff of a marine pond would scarcely equal. Nor was there wanting a section of Tolosan cheese, nor a four-year-old chaplet of black pennyroyal, and ropes shorn of their garlic and onions, nor your mother's pot of foul resin, the depilatory of dames.[24]

Fragrance in the home, however, was a matter of practical house-keeping for the ancients, as well as aesthetic pleasure. Clothes stored in cedar chests, for example, were not only kept fragrant, but also protected from moths, which dislike the scent of cedar. Likewise, incense burning in storerooms both perfumed the wares within and helped keep out rodents. Incense was also believed to purify the air of the contaminating influences of illness or misfortune. Thus, even poorer families would keep burning censers at their front doors in order to protect their homes from the malignant emanations of the world outside. For those people who lived in apartments, a pot of fragrant violets on a window-sill might serve the purpose instead.[25]

Perfumes were also of use in masking unpleasant odours in the home, such as those arising from waste products. Lamp oil, which had a notoriously bad smell, was perfumed by those who could afford it, adding scent to light.[26] Similarly, incensing clothes with perfume would disguise any unpleasant odours inherent in the cloth or dyes employed in their manufacture. Purple dye, for example, used only in the clothing of the royalty and aristocracy because of its high cost, had the pungent odour of the decayed shellfish from which it was made. Thus Martial, wishing to call up the image of a particularly potent stench, writes of a fleece (which would be malodorous in itself) that has been 'twice dyed in purple'.[27] In another epigram he mocks a woman who constantly wears the high-status purple by saying it is the foul smell she cares for rather than the colour: 'Because Philaenis night and day wears garments dipped in every kind of purple, she is not ambitious or proud. She is pleased with the smell, not with the hue.'[28]

Whatever the reasons for using perfume in the home, the effect was the creation of an atmosphere redolent of aromatics, which formed as integral a part of the house as its furniture or painted mosaics. Such homes, at their best, would be oases of fragrance in the city, enveloping their inhabitants and guests in a sweet, inviting blanket of scent.

THE PERFUMED BANQUET

> And always at the banquet crown your head
> With flowing wreaths of varied scent and hue
> Culling the treasures of the happy earth;
> And steep your hair in rich and reeking odours,
> And all day long pour holy frankincense
> And myrrh, the fragrant fruit of Syria,
> On the slow slumb'ring ashes of the fire.[29]

A feast in antiquity was not a proper feast without its complement of perfume, as the above lines of Greek verse attest. This extract also tells us the three kinds of scent necessary for a successful dinner party in ancient Greece or Rome: the fragrance of fresh flowers, perfumed unguents and incense.

Fresh flowers were often strewn on the floor of the room in which the banquet was held. At one luxurious feast, the floor was said to be covered so thickly with different blossoms that it resembled a 'most divine meadow.'[30] The table, in turn, might be made ready by being rubbed with mint leaves.[31] The diners would themselves be adorned with fragrant garlands. These garlands could be made of many different kinds of flowers or leaves, such as roses, violets, hyacinths, apple-blossoms, thyme, rosemary, myrtle, bay and parsley. Worn around the forehead as a crown, a garland was supposed to alleviate the effects of drinking, and worn on the breast it was said to enliven the heart. In an olfactory example of gilding the lily, perfumes might be added to the wreaths to make them more odorous.

Scented water would be offered to the guests in between courses for washing their sticky hands, as most foods were eaten with the fingers. The diners would be further scented with perfumed unguents brought to them in alabaster boxes by slaves. Innovative hosts would try and introduce an element of novelty into this arrangement, as recorded in the following description:

> For he did use no alabaster box
> From which t'anoint himself; for this is but
> An ordinary, and quite old-fashioned thing.
> But he let loose four doves all dipp'd in unguents,
> Not of one kind, but each in a different sort;
> And then they flew around, and hovering o'er us,
> Besprinkled all our clothes and tablecloths.[32]

At the luxurious banquet depicted by the first-century Roman satirist Petronius, the host has a hoop which descends from the ceiling with boxes of perfume for the guests.[33] Nero, in turn, had pipes installed in his dining-room ceiling to sprinkle fragrant water on his guests, while a later emperor, Elagabulus, reputedly covered his guests with such an abundant floral shower that some of them suffocated.[34]

On even relatively informal occasions, perfume would be offered to a guest as naturally as we offer a visitor a cup of coffee. In Plautus's play *Mostellaria*, for example, a woman says to her lover who has dropped in for a drink and a game of dice: 'Come, sit down. Boy! Bring in some water for our hands. Where are the dice? Will you have some perfume?'[35] At a lavish dinner party the perfumes offered could be very costly and so afford less wealthy or extravagant guests a rare pleasure. Indeed so much emphasis was placed on the enjoyment of sweet scents at dinners, that occasionally the food itself seemed paltry by comparison. Martial writes sarcastically of one such occasion: 'Good unguent, I allow, you gave your guests yesterday, but you carved nothing. 'Tis a droll thing to be scented and to starve.'[36]

The food served at dinner would, of course, while appealing to the sense of taste, also add its quota of scent to the ambience. 'All the house with the rich odour steamed', is how one classical writer describes the olfactory effect of a sumptuous feast.[37] Ancient cooks are represented as very proud of the rich fragrance of the dishes they prepare. Thus one is said to boast that the smell of his food is enough to bring a man back from the dead.[38] Another remarks that the odour of his dishes is so compelling as to have the same mesmerizing effect upon those who smell it as sirens have upon those who hear them. In the latter case, the only way to escape the spell of the sweet-voiced sirens is to plug up one's ears. In the former:

> Once I have properly arranged my kitchen . . .
> Such shall be the savoury smell, that none
> Shall bring themselves to pass this narrow passage;
> And every one who passes by the door
> Shall stand agape, fix'd to the spot, and mute,
> Till some one of his friends, who's got a cold
> And lost his smell, drags him away by force.[39]

The ancients, in general, were quite fond of strong-flavoured and smelling foods, such as cheese, garlic and onion. In fact, one favourite Roman seasoning agent was a fermented fish sauce, *garum*, which had quite a putrid smell. Meals – which could be comprised of many courses interspersed with entertainments – customarily ended with a sweet-scented serving of fruit. The Roman writer Juvenal, for example, speaking of a dinner host, writes that for dessert 'he will order fruits to be served whose scent alone would be a feast.'[40]

Not content with the natural odours of foods, however, the ancients often added perfume to their dishes, as in the following excerpt from Athenaeus' *Banquet of the Learned*, in which a cook describes one of his confections.

> 'I call this dish the Dish of Roses. And it is prepared in such a way, that you may not only have the ornament of a garland on your head, but also in yourself, and so feast your whole

body with a luxurious banquet. Having pounded a quantity of the most fragrant roses in a mortar, I put in the brains of birds and pigs boiled and thoroughly cleansed of all the sinews, and also yolks of eggs, and with them oil, and pickle juice, and pepper, and wine . . .' And while saying this, he uncovered the dish, and diffused such a sweet perfume over the whole party, that one of the guests present said with great truth –

The winds perfumed the balmy gale convey
Through heav'n, through earth, and all the aerial way

– so excessive was the fragrance which was diffused from the roses.[41]

Such intermingling of odours made it at times impossible to distinguish the product of a cook from that of a perfumer. A character in a play by Cratinus, for example, proclaims:

Consider now, how sweet the earth doth smell.
How fragrantly the smoke ascends to heaven:
There lives, I fancy, here within this cave
Some perfume seller, or Sicilian cook.[42]

A perfume might well add a pleasant flavour to food, as in the case of attar of roses. Yet, if it were bitter in flavour, as myrrh, for example, was, it could instead detract from the taste of the dish. In such cases, however, scent seems to have taken precedence over flavour. Thus Pliny writes that 'some people actually put scent in their drinks and it is worth the bitter flavour for their body to enjoy the lavish scent both inside and outside'.[43] Myrrh, in fact, was a popular ingredient in wines, along with the essences of various flowers and scented honey. In one Greek play, Bacchus, god of wine, describes his favourite wine in terms of its floral bouquet:

violets and roses mix their lovely scent
And hyacinths, in one rich fragrance blent.[44]

In the Roman play *Curculio* by Plautus, a wine lover refers to wine as 'my myrrh, my cinnamon, my rose, my saffron, my cassia, my fenugreek.'[45]

If perfumes were often added to wine, wine was also added to perfumes, contributing a pleasant fragrance of grapes to the compound. Honey, which would have different scents according to the flowers used by the bees in its confection, was another common ingredient of perfumes. In the modern West we think of perfume and food as constituting two very different categories, distinct both in odour and in edibility. In the ancient world, however, there was no such division: foods could be perfumed and perfumes could be, and were at times, eaten. Whereas most modern perfumes would be highly distasteful and probably poisonous, an ancient perfume composed of, say, attar of roses, cinnamon, honey and wine could be quite delightful to the palate. This confusion of the olfactory and the gustatory is described by Juvenal, who writes of a guest at a banquet who 'pours foaming unguents into her [wine], and drinks out of perfume flasks, while the roof spins dizzily round, the table dances, and every light shows double!'[46] The different scents enjoyed at a banquet – perfume, flowers, incense, food and wine – therefore, would all be variations on an olfactory theme. The following verse by Xenophanes offers an idea of this odorous interplay.

> A willing youth presents to each in turn
> A sweet and costly perfume . . . another pours out wine
> Of most delicious flavour, breathing round
> Fragrance of flowers, and honey newly made,
> So grateful to the sense, that none refuse,
> While odoriferous gums fill all the room.[47]

Ancient banquets were carefully organized so as to provide for the pleasure, and indeed surfeit, of all the senses. The guests at a banquet would eat comfortably reclined on couches. (In fact, to sit down to eat was regarded as such hardship that one upper-class Roman, who disapproved of Julius Caesar, gained a reputation as an ascetic for vowing to eat seated so long as Caesar remained in power.)[48] Music would usually be performed during or after the dinner, while other entertainment, such as dancers or jugglers, might be provided in between courses. Incense was customarily burnt at the end of the meal, if not before, as the guests enjoyed themselves with goblets of wine and discussed the issues of the day. This incense served not only to clear away the scents of the food and render the atmosphere agreeably spicy, but also as an offering to the household gods who would be invisibly present during every meal.[49]

If the enjoyment of wine was too free and the post-dinner discussion too heated, however, the evening might well end on a disagreeable note. The comic poet Alexis writes that 'many a banquet which endures too long' ends in 'blows and drunken riot'.[50] In one of his plays, Sophocles tells of an angry dinner guest who goes so far as to throw his chamber-pot at another guest's head:

> He in his anger threw so well
> The vessel with the evil smell
> Against my head, and fill'd the room
> With something not much like perfume;
> So that I swear I nearly fainted
> With the foul steam the vessel vented.[51]

Certainly an olfactory anticlimax to the aromatic elaboration of the evening!

The majority of the inhabitants of the ancient world, of course, enjoyed much simpler meals than the banquet described above. This was mostly due to a lack of wherewithal, but also, as in the case of the vegetarian Pythagoreans – out of concerns of ethics or health. In such cases, a meal might consist only of 'a simple loaf ... and a pure cup of water'.[52] The ancients, however, were well able to appreciate the fragrance of even the simplest of foods. Thus Nicostratus writes of a freshly baked bread:

> A large white loaf. It was so deep, its top
> Rose like a tower quite above its basket.
> Its smell, when the top was lifted up,
> Rose up, a fragrance not unmix'd with honey
> Most grateful to our nostrils, still being hot.[53]

Even water, which we tend to think of as tasteless and odourless, was appreciated for its savour by the ancients. One classical writer, for example, praises a fountain in Boetia, saying it produces 'ambrosial water, like fresh honey sweet'.[54] Another notes that the water of the Hyannis river, 'at a distance of five days' journey from its head, is thin and sweet to the taste'; but that 'four days' journey further on it becomes bitter'.[55] Yet another remarks with authority that water from springs flowing north-east 'must inevitably be clear, fragrant and light'.[56]

Thus, in appreciation of the simpler pleasures of ancient life, we will end our discussion of the 'perfumed banquet' with a description of a plain but aromatic Pythagorean feast.

- The banquet shall be figs and grapes and cheese,
  For these the victims are which the strict law
  Allows Pythagoras' sect to sacrifice
- By Jove, as fine a sacrifice as any.[57]

POMP AND PERFUME

Fragrance was an important element of public entertainment in antiquity. Roman theatres, for example, customarily had their stages sprinkled with saffron or other scents.[58] The amphitheatres of Rome, in turn, were sweetened by fountains spraying perfumed water into the air. Such scents helped mask the many unpleasant odours which often accompanied public spectacles: from the mal-odours of the crowd to the smell of blood arising from the massive slaughter of wild animals in the arena (as many as 5,000 in one day) to the stench of Christians burning as torches.[59]

For important gatherings, such as the holding of athletic games, the use of perfumes could be quite lavish. This can be seen in the description of the games held by Antiochus Epiphanes, the king of Syria, in the second century BC in the city of Daphne. As part of the parade which initiated the games, two hundred women sprinkled every one with perfumes out of golden pitchers. During the games themselves,

> on the first five days every one who came into the gymnasia was anointed with a saffron perfume shed upon him out of golden dishes ... And in a similar manner in the five next days there was brought in essence of fenugreek, and of amaracus, and of lilies, all differing in their scent.[60]

Similarly, the parade held in Alexandria by the Egyptian king Ptolemy Philadelphus in the third century BC is said to have boasted among its extravagances boys in purple tunics carrying frankincense, myrrh and saffron on golden dishes, a giant figure of Bacchus pouring out libations of wine, golden-winged images of Victory bearing incense burners, camels loaded down with spices, and innumerable floral decorations. The abundance of these last was 'quite incredible to the foreigners' given that the festivities took place in the middle of winter.[61]

Putting on a good show in antiquity, therefore, involved putting out a good scent. The spicy, sweet scents offered to the spectators at such events would not only serve to please and excite them, but would help make them feel involved in the activities in a way that a purely visual display could not. Not only would the spectators see and hear the pageantry, they would breathe it in and feel identified with it and each other.

In the modern West, we tend to think of the use of perfumes as a purely individual matter. In antiquity, however, collective perfuming was an important means of entertaining and impressing the masses and of establishing group solidarity.

THE FRAGRANCE OF KISSES

Breath of balm from phials of yesterday, of the last effluence that falls from a curving jet of saffron; perfume of apples ripening in their winter chest, of fields lavish with the leafage of spring; of Augusta's silken robes from Palatine presses, of amber warmed by a maiden's hand, of a garden that stays therein Sicilian bees; the scent of Cosmus' alabaster boxes, and of the altars of the gods; of a chaplet fallen but now from a rich man's locks – why should I speak of each? Not enough are they; mix them all; such is the fragrance of my boy's kisses at morn.[62]

Nowhere is there a better catalogue of fragrance in ancient literature than in Martial's evocations of the aroma of kisses. 'Breath of a young maid as she bites an apple', he writes in another example,

perfume such as when the blossoming vine blooms with early clusters; the scent of grass which a sheep has just cropped; the odour of myrtle, of the Arab spice-gatherer, of rubbed amber; of a fire made pallid with Eastern frankincense; of the earth when lightly sprinkled with summer rain.[63]

A whole world of fragrance is called up by Martial's lines – from the sweet, tangy scent of apples to the heady perfume of ointment in an alabaster box, from the fresh aroma of cut grass to the spicy smoke of incense burning in the fireplace – all in response to the image of a kiss.

Martial's association of fragrance and kisses was part of a whole ancient olfactory/amatory complex in which personal attraction was conveyed in terms of sweet scents. This complex is expressed, for example, in the following Greek epigram: 'I send thee sweet perfume, ministering to scent with scent.'[64] Perfume here constitutes a love offering to a beloved who is herself a perfume.[65]

The association of love with the sense of smell is also made by the Roman playwright Plautus in *Miles Gloriosus*. In this play a courtesan, Acroteleutium, convinces a man, Pyrgopolynices, of her love by pretending within his earshot that she can sense his presence by his odour.

ACROTELEUTIUM: ... the man I want is not inside.

MILPHIDIPPA: How do you know?

ACROTELEUTIUM: My sense of smell tells me; if he were inside, my nose would sense it from the odour.

PYRGOPOLYNICES [*aside to Palaestrio*]: She's a diviner. She's in love with me, and that's why Venus has given her powers of prophecy.

ACROTELEUTIUM: The man I want to see is around here somewhere; I can certainly smell him.

PYRGOPOLYNICES [*aside to Palaestrio*]: Gad! This woman sees more with her nose than she does with her eyes.

PALAESTRIO [*grinning*]: That's because she's blind with love, sir.[66]

In order to achieve the ideal of the sweet-scented lover, perfumes were used by both men and women in the ancient world. 'Now that I'm in love with Casina ... I keep the perfumers busy; I use all the nicest ointments to please her', remarks an enamoured Athenian in Plautus's *Casina*.[67] Similarly, in a play by Archilochus, a woman is described as

Displaying hair and breast perfumed
So that a man, though old
might fall in love with her.[68]

With regard to royal lovers, the third century BC king of Macedonia, Demetrius Poliorcetes, is said to have exhausted his stock of perfumes trying to find a scent which would render him attractive to a certain female flute player.[69] When in the first century BC Cleopatra, dressed up as Venus, sailed to the seduction of Mark Antony, 'an indescribably rich perfume, exhaled from innumerable censers, was wafted from the vessel to the river banks'.[70] Furthermore, at one of the banquets Cleopatra gave for Antony she filled a room with a foot and a half of roses.[71]

As a fragrant token of their adoration, lovers would present their loved ones with floral crowns. In the following epigram, a lover describes how he will use a variety of scented flowers to maximize the olfactory and visual beauty of the crown and make it a fit adornment for his beloved.

> I will plait in white violets and tender narcissus mid myrtle berries, I will plait laughing lilies too and sweet crocus and purple hyacinths and the roses that take joy in love, so that the wreath set on Heliodora's brow, Heliodora with the scented curls, may scatter flowers on her lovely hair.[72]

Such floral crowns might also be placed by lovers on the door-steps of their loved ones' homes. Lucretius speaks of 'the tearful lover [who] smothers the thresh-hold with flowers and garlands, and anoints the haughty door-posts with marjoram'.[73] In these cases the wreaths would serve as a public, as well as private, display of devotion, marking by sight and scent a house as the home of one who was loved.

Apart from their purely romantic uses, wreaths were customary offerings to make to a god, particularly the gods of love. A jealous lover, for instance, says of a courtesan in Plautus's *Asinaria*:

> If she bids her maidens take to Venus
> Or Cupid garlands, wreaths, or precious ointments,
> Your slave shall see that Venus gets them all,
> And not some man to whom she's sending tokens.[74]

Given the sacred associations of the act, therefore, placing a floral crown on the door of one's beloved would turn the house into something of a shrine to love. Athenaeus, pondering this matter, writes:

> Perhaps the offering of the crowns is made, not to the beloved object, but to the god Love. For thinking the beloved object the statue, as it were, of Love, and his house the temple of Love, they, under this idea, adorn with crowns the vestibules of those whom they love.[75]

The association of love with fragrance was not merely rhetorical, for love was personified in the fragrant persons of Aphrodite and Eros, or Venus and Cupid as they were called by the Romans. These gods were not only sweet-scented themselves, but also delighted in the presence of perfumes. Thus Plato writes of Eros:

'Love will not settle on body or soul or aught else that is flower-less or whose flower has faded away; while he has only to light on a plot of sweet blossoms and scents to settle there and stay.'[76] When lovers adorned themselves with perfumes, therefore, the sweet scents served not only to attract their loved ones, but Love itself.

ON BEING IN ILL ODOUR

Thais smells worse than a grasping fuller's long-used crack [a pot filled with urine], and that, too, just smashed in the middle of the street; than a he-goat fresh from his amours; than the breath of a lion; than a hide dragged from a dog beyond Tiber; than a chicken when it rots in an abortive egg; than a two-eared jar poisoned by putrid fish-sauce. In order craftily to substitute for such a reek another odour, whenever she strips and enters the bath she is green with depilatory, or is hidden behind a plaster of chalk and vinegar, or is covered with three or four layers of sticky bean-flour. When she imagines that by a thousand dodges she is quite safe, Thais, do what she will, smells of Thais.[77]

Just as the ancients were outspoken in their praise of sweet scents, so they were in their denunciation of foul odours.[78] Martial's description of the bad smell of Thais, a woman who had evidently got on his bad side, contains some of the most potent olfactory images ever recorded. A fuller's jar of stale urine or a rotten egg, for example, are pungent enough images in themselves. Martial, however, enhances them by having the jar smashed in the middle of a street where it would spread its odour far and wide, the egg stinking not only of itself but of a dead chick within. Thais, who reputedly smells worse than all these things, cannot even mask her stench with the strong odours of beauty preparations.[79]

Personal odour, thus, was of great concern to the ancients. Aristotle, indeed, devotes a section of his *Problemata* to the subject of foul body odour. 'Why has the armpit a more unpleasant odour than any other part of the body?', he queries, and 'why is it that those who have a rank odour are more unpleasant when they anoint themselves with unguents?'[80]

A particular focus of this concern with body odour was breath, for just as a fragrant kiss was a romantic ideal, so was foul breath a subject of disgust and ridicule. Martial writes bluntly that 'when gluttonous Sabidius blows on a hot tart to cool it off it turns into excrement'.[81] The epigrammatist Lucilius says of Telesilla's breath that it is worse than all the famous stinks of mythology: 'Not Homer's Chimaera breathed such foul breath, not the fire-breathing herd of bulls of which they tell, not all Lemnos [cursed with a foul odour by Venus] nor the excrements of the Harpies, nor Philoctetes' putrefying foot.'[82]

The main causes ascribed to bad breath in the ancient world were eating pungent foods and drinking wine. Garlic and onions were particularly cited as culprits and lovers were supposed to refrain from eating these foods in order to keep their kisses sweet. Martial writes of leeks in the same vein, saying that after eating them, you should 'give kisses with shut mouth'.[83] There are many images in ancient literature of the stench of drunkenness. Juvenal has a powerful description of a rich drunken woman who vomits until 'the gilt basin reeks of Falernian [wine]'.[84] Martial, in turn, writes sarcastically that: 'He who fancies that Acerra reeks of yesterday's wine is wrong. Acerra always drinks till daylight.'[85]

If indulgence in food and wine could cause bad breath, so could abstaining from food and drink, for the mouth grows stale after long periods without eating. One instance of when this happens is during a night's sleep. Another instance is during fasts. Fasting was widely practised in the ancient world, both as a health measure and as a religious rite, and the malodorous breath which resulted from it was a fairly common phenomenon. Martial, for example, in one of his condemnations by smell, compares a woman's foul odour to 'the breath of fasting Sabbatarian Jews'.[86] Aristotle was himself intrigued by the phenomenon. 'Why is it that the mouths of those who have eaten nothing, but are fasting, have a [strong] odour . . .?', he asks in his *Problemata*.[87]

Other body odours also came in for their share of criticism by the ancients. The odour of stale perspiration, for example, was often described as similar to the smell of a goat. Martial writes that Thais smells worse than a 'he-goat fresh from his amours', and that with the arrival of male puberty there 'comes a goatish odour'.[88] Aristophanes, similarly, describes a character in *The Archarnians* as having 'arm-pits stinking as foul as a goat'.[89]

The ancients employed a variety of techniques to prevent and disguise the body odours described above. Perfumed pastilles were available, for example, as a remedy for bad breath. Martial writes mockingly of a woman who tries to mask her alcoholic breath by devouring pastilles made by the famous Roman perfumer, Cosmus.

> That you may not smell strong of yesterday's wine, Fescennia, you devour immoderately Cosmus's pastilles. That snack discolours your teeth, but is no preventive when an eructation returns from your abysmal depths. What if the stench is stronger when mixed with drugs, and redoubled the reek of your breath carries farther? So away with tricks too well known, and detected dodges, and be just simply drunk![90]

Those who could not afford such preparations could instead chew on scented leaves or berries. Pliny recommended myrtle berries for this purpose.[91] Martial, in another of his epigrams, says that 'Myrtale is wont to reek with much wine, but to mislead us, she devours laurel leaves'.[92] Laurel leaves were chewed by the priestesses of Apollo at Delphi to acquire inspiration. In Myrtale's case, however, they are a sign rather of intoxication. A character in Aristophanes' *Thesmophoriazusae* says that women chew garlic in the morning to hide the scent of a night's drinking – an effective but not very aesthetic remedy.[93] Paradoxically, Pliny recommended a drink of wine at bedtime to prevent bad breath in the morning.[94]

The odour of perspiration was dealt with in various fashions as well. Bathing was the most important, and in the Roman empire frequent baths, as many as three daily, were an established institution. The Romans and Greeks, both men and women, also removed their underarm hair, which would help prevent underarm odour. Besides, the Romans made use of aluminium salts, the main ingredient in modern antiperspirants, to check perspiration.

A good, strong perfume, of course, could serve to hide a multitude of unwanted odours, and perfumes were undoubtedly often used for this purpose. Such olfactory cover-ups only worked up to a point, however, for cynics were always ready to suspect the overly perfumed of underlying ill odour. 'He is not well-scented who is always well-scented,' huffs Martial of one lavishly perfumed fellow.[95] In Plautus's play *Mostellaria*, it is said of older women who cover themselves with cosmetics and scent: 'When their perspiration mixes with the perfume, it's just as if a cook had poured all his soups together. You'd never know what it is they smell like – but the stench is awful.'[96] Not even perfumes then, could ultimately ensure one a state of olfactory grace. Martial thus harshly observes to one scented lady: 'Whenever you come we fancy Cosmus [the perfumer] is on the move, and that oil of cinnamon flows streaming from a shaken out glass bottle. I would not have you Gellia, pride yourself upon alien trumpery. You know, I think, my dog can smell sweet in the same way.'[97]

ODOUR CLASSES

Odours were not simply a matter of aesthetic preference in the ancient world, but also a means by which different classes of people were categorized.[98] The most obvious class division of this sort was that between rich and the poor. The rich, of course, could afford all the olfactory niceties which the poor could not: perfumes and incense, scented lamp oil, gardens, well-ventilated homes kept clean and sweet by slaves. By contrast we have Martial's olfactory portrait of a poor family's home: urine leaking from a cracked chamber-pot, a jar stinking of fish, another full of foul resin, all intermingled with the pungent odours of garlic and onions.

Indeed, the very scent of perfume indicated wealth, for only the rich could afford to buy perfumes. The story is told of the well-perfumed Syrian king Antiochus Epiphanes, that one day while at the baths, a man approached him and said 'You are a happy man, O king; you smell in a most costly manner.' The king, pleased with this compliment, ordered a pitcher of perfumed unguent to be poured over the man's head. The unguent spilled over the floor and the poor people in the vicinity hastened to scoop up some of this olfactory wealth for themselves.[99]

Paradoxically, the scent of money itself could indicate poverty, for poor people had the custom of carrying coins in their mouths – a practice which gave a metallic smell to their breath.[100] In general, however, money was too highly valued for any ill odour to be attached to it. For example, the following anecdote was told of the first-century emperor Vespasian. In response to a complaint by his son Titus about the taxing of public urinals, Vespasian handed the youth a coin and enquired if it smelled bad. When Titus replied in the negative, Vespasian said: 'That's odd: it comes straight from the urinal!'[101]

Other olfactory distinctions, besides that of rich and poor, were also made in antiquity. Among the working classes, certain trades – tanner, fishmonger, fuller, for example – were characterized as foul due to the odours associated with them. Martial speaks in disgust, for instance, of having to endure the embraces of such tradespeople after coming back from a journey: 'Upon you all the neighbourhood presses, upon you the bristly farmer with a kiss like a he-goat, on this side the weaver crowds you, on that the fuller, on this the cobbler who has just been kissing his hide.'[102]

Another important olfactory division was made between city and country dwellers. On the one hand, of course, the countryside would naturally tend to be more fragrant than the crowded, dirty urban landscape. City dwellers, however, were inclined to characterize their rustic counterparts as uncouth bumpkins stinking of goats and garlic. Martial speaks above of a 'bristly farmer with a kiss like a he-goat'. In *Clouds*, Aristophanes has a character define the differences between himself and his wife in terms of their different social and olfactory categories: 'I belonged to the country, she was from the town ... On the nuptial day, when I lay beside her, I was reeking of the dregs of the wine-cup, of cheese and of wool; she was redolent with essences, saffron, voluptuous kisses, the love of spending.'[103] The following interchange from Plautus's play *Mostellaria* between a city slave called Tranio and the country slave Grumio indicates that these olfactory distinctions between rich and poor, city and country, held even among slaves.

> TRANIO: You smell of garlic. You thing of filth, you hick, goat, pig-sty, you mud-and-manure, you!
>
> GRUMIO: Well, what do you want? Everybody can't smell of [fine] perfumes just because you do, or sit at the head of the table, or live on the fine food you do. You can have your grouse and fancy fish and fowl. But let me go my way on a meal of garlic. You're rich and I'm poor, and that's that.[104]

Interestingly, Socrates is said to have argued against the use of perfumes because they disguised the olfactory differences between freeborn citizens and slaves:

> If you perfume a slave and a freeman, the difference of their birth produces none in the smell; and the scent is perceived as soon in the one as in the other; but the odour of honourable

toil, as it is acquired with great pains and application, so it is ever sweet and worthy of a [freeman].[105]

It is evident that the olfactory class distinctions of antiquity were not simply based on actual differences in odour, but were also symbolic in nature. The wealthy, for example, were categorized as fragrant not only because of their use of perfumes, but because of their high status in society, while the poor were characterized as foul not only because of the malodours of their impoverished living conditions, but because of their low social status. The ancients themselves were aware, to some extent, of the operation of these prejudices. Consider, for example, the following extract from a play by Pherecrates in which flatterers praise the sweet scent of a wealthy man:

O you who sigh like mallows soft,
Whose breath like hyacinths smells,
Who like the melilotus speak,
And smile as doth the rose,
Whose kisses are as marjoram sweet,
Whose action crisp as parsley.[106]

Similarly, a character in a verse by Diodourus complains that even a foul-smelling plutocrat will be said by sycophants to be fragrant:

So that if any one should eat a radish,
Or stinking shad, they'd take their oaths at once
That he had eaten lilies, roses, violets;
And that if any odious smell should rise,
They'd ask where you did get such lovely scents.[107]

One can imagine that the reverse was also true, that even a well-scented poor man would be said to be ill-odoured by those who held him in contempt because of his social status. Being in metaphorical good odour in the ancient world, therefore, depended on more than simply being fragrant.

Among the different classes of people who were categorized by odour were women and men. There are several references in ancient literature to the different characteristic odours of the sexes. In Aristophanes' *Lysistrata*, for example, there is a scene in which men and women are arguing over which sex is superior. The men respond to the perceived threat to their authority by taking off their tunics, 'for a man must savour of manhood'.[108] The women, in turn, take off *their* tunics, for 'women must smell the smell of women in the throes of passion.'[109] The battle between the sexes is thus presented as an olfactory contest. As to the outcome, however, we do not know, for the debate is interrupted.

Socrates held that 'as there is one sort of dress fit for women and another for men, so there is one kind of smell fit for women and another for men.'[110] Although women and men were acknowledged to have different natural scents, when it came to perfumes they generally wore the same scents – attar of roses, oil of cinnamon, myrrh, spikenard – as there were no specifically masculine or feminine perfumes. There *was* a feeling in certain ascetic male quarters, however, that perfume should be worn only by frivolous females and that its use by men was a sign of effeminacy.[111] Socrates was of the opinion, for instance, that the only legitimate scent for a man was the smell of the oil worn in gymnastics.[112]

There are more references to the odour of women in ancient literature than there are to that of men. The basic olfactory classification made of women was to associate desirable women with fragrance and undesirable women with stench. Attractive young women, thus, are constantly described in terms of sweet scents in ancient literature. To enhance their desirable sweetness, as well as to purify themselves, brides would customarily anoint themselves with perfumes for their weddings. After some years of marriage, however, a woman might find her status changed from desirable to undesirable and from fragrant to foul. Thus in Plautus's *Asinaria*, an unhappily married man says that, while the kisses of a courtesan are sweet, he'd rather drink bilge-water than kiss his wife.[113]

Older women, as irremediably past the age of desirability, were particularly associated with bad smells. In *Mostellaria* the mixed smell of perspiration and perfume of 'old hags' is said to be indescribably awful. Martial describes a bony, wrinkled old woman as having the odour of a goat.[114] Horace writes of an older woman who sends him unwanted love-letters: 'What sweatiness, and how rank an odour everywhere rises from her withered limbs!'[115] It *was* considered possible for the rare older woman to be attractive, and therefore fragrant, however, as the following Greek epigram acknowledges: 'Charito has completed sixty years, but still ... her skin without a wrinkle distils ambrosia, distils fascination and ten thousand graces.'[116]

Prostitutes constituted another group of women singled out as bad-smelling by the ancients. Juvenal writes of 'a brothel reeking with long-used coverlets', and a 'strumpet that stands naked in a reeking archway'.[117] A promiscuous woman is described by him as carrying home with her 'all the odours of the stews'.[118] A character in one of Plautus's plays exclaims: 'You surely don't want to mingle there with those common prostitutes ... plastered with poor perfumery, that smell of pothouse and profession.'[119] Lucilius, in turn, states that old prostitutes smell like swine.[120] Timocles wrote that it was possible to distinguish a 'good' girl from a prostitute by their different scents, that of the former, of course, being preferable to that of the latter.[121]

The foul smell accorded to prostitutes was indicative both of the filthiness of the conditions in which they often worked, and of their very low social status. Prostitutes were available women, but they were not usually considered very desirable. Not only were prostitutes 'cheap goods', their dissoluteness made them disruptive to the social order. This latter quality gave an added pungency to their stench, for the stink of the brothel was also the metaphorical stench of a corrupted social body.

A more attractive or well-born prostitute might occupy the higher rank of a courtesan, and thus be deemed fragrant. None the less, the higher-status courtesans were, like prostitutes, considered disruptive of family life and social order. Therefore, they are often described as agents of discord in ancient plays and stories. While they were supposed fragrant because of their attractiveness, it was a dangerous sweetness, intoxicating but potentially ruinous. The seductive, perfumed Cleopatra, for example, is portrayed as ultimately leading Mark Antony to his downfall.[122]

These olfactory stereotypes of women are also present in ancient mythology. Circe, with her potions and perfumes, is an example of the fragrant seductress. Fragrant virtue is represented in such feminine types as the flower-garlanded Graces. The plight befalling women who rebel against the established order is described in the story of the women of Lemnos who, having failed to make the proper offerings to Venus, goddess of fragrance, were cursed with a foul odour. Witches, as completely antagonistic to the social order, are even more repulsive: 'Haggard and loathly with age is the face of the witch . . . her breath poisons air that before was harmless,' writes the Roman poet Lucan.[123] The Harpies, excrement-dropping bird-women, are particularly potent models of foul womanhood which even today are used to characterize 'shrewish' women.

While different types of womanhood were represented as fragrant or foul according to these models, however, to a certain extent *all* women were thought to be foul-smelling in the ancient world. In *On the Nature of Things*, for example, Lucretius writes that even the most beautiful of women 'reeks of noisome smells' in private.[124] This underlying 'foulness' of women was expressed in the ancient association of women with the moon, which, in turn, was associated with corruption.[125] Men, on the other hand, were linked with the sun, considered productive of sweet scents. From this perspective, the tradition of perfuming brides in the ancient world could be understood in part as a kind of cultural processing whereby naturally foul, disruptive women were symbolically turned into sweet, obedient helpmates.

Olfactory symbolism, thus, was used very effectively to pass value judgements on different groups of people in antiquity. Given the strong emotional and physical reactions of pleasure or disgust which smells inspire, such an olfactory classificatory system would have been a potent aid to maintaining different classes in their 'proper' place in the social order.

THE SMELL OF THE BATTLE

Military life was characterized by a variety of different odours in antiquity. Harsh odours – the stench of the wounded and dead, the acrid smoke of burning fields and towns – were an intrinsic part of ancient battles.[126] Soldiers themselves, sweaty in hot armour, reeking of their rations of cheese and onions, were often caricatured as foul-smelling. Martial, for example, gives the smell of a well-worn army boot as one of the world's worst stenches.[127] For Aristophanes the characteristic smell of a soldier is that of onions emanating from his knapsack.[128]

Freedom from military duty, on the other hand, is called a 'gentle fragrance' by Aristophanes.[129] Similarly, he has a passage in *The Archarnians* in which truces of different periods of length are judged by their odours. A three-year truce, which gives the enemy time to rebuild their fleet, is said to smell of pitch and ships. A ten-year truce, during which the opposing parties can seek military alliances, 'smells strongly of the delegates, who go around the towns to chide the allies for their slowness'. A thirty-year truce, however, long enough to assure freedom from the pressures of war, 'has the aroma of nectar and ambrosia'.[130]

This passage is not as purely metaphorical as it sounds. Interestingly, whereas we associate military surrender with a white flag, in some parts of the ancient world the leader of a besieged town would indicate a surrender by holding out an incense burner over the city walls.[131] The besieging army camped outside would thus get wind of their triumph through their noses – a sweet victory indeed!

Incense had other military uses as well. It was, of course, of the utmost importance as a means of petitioning the gods for victory. While attracting the gods, it could also be employed to keep the army camps free of snakes, as described in the following lines from *The Civil War* by the first-century Roman writer, Lucan.

> The limits of the camp were surrounded by a fire of fumigation, in which elder-wood crackled and foreign galbanum bubbled; the tamarisk of scanty leaf, Eastern *costos*, powerful all-heal, Thessalian centaury, fennel, and Sicilian *thapsos* made a noise in the flame; and the natives also burned larchwood, and southernwood whose smoke snakes loathe.[132]

Furthermore, incense was sometimes used to purify a conquered town or a battle site of the contaminating odours of the blood of the enemy.[133]

Ancient warfare was thus not all 'foul'. In fact, in certain cases, the claim by revellers in an ancient poem that 'ointments and perfumes [shall be] our war-cry fierce'[134] was almost appropriate. Military expeditions to the scented lands of the East inevitably left an olfactory mark on the Greek and Roman armies. Alexander the Great, for example, is said to have come to enjoy perfumes so much after his contact with the aromatic traditions of the Orient, that he had his rooms sprinkled with rich scents wherever he stayed.[135] Indeed, perhaps because of his use of fragrance, Alexander is described as having exuded an odour so sweet that his clothes were all impregnated with it.[136]

In the Roman army, shields, spears and standards would be anointed with perfume on holidays. Perfumes were also used for personal adornment by those Roman soldiers who could afford them. Pliny, for example, tells of warriors who had perfumed hair under their helmets.[137] Julius Caesar reportedly boasted of his soldiers that 'My men fight just as well when they are stinking of perfume.'[138] On the other hand, he once had to chide some of his companions-in-arms for complaining that their asparagus had been served with myrrh rather than olive oil.[139]

Better than perfume for a Roman commander, however, was the scent of one of the leafy crowns traditionally awarded to victors in battle. The liberator of a besieged army, for example, was given a crown of grass and wildflowers, taken from the place in which the army was enclosed. A crown of myrtle was presented to a commander who had won a battle.[140] This custom is alluded to in the following Greek verse:

> I'll wreathe my sword in myrtle bough,
> The sword that laid the tyrant low,
> When patriots, burning to be free,
> To Athens give equality.[141]

A general responsible for a major victory wore a crown of laurel leaves during his triumphal march through Rome. Even today a laurel crown stands as a symbol of victory, though for most of us it is a purely visual image with no olfactory connotations. For an ancient Roman military leader, however, the odour of a crown of fragrant laurel leaves would be the ultimate smell of success.

HEALING BY SMELL

Scents, either inhaled through the nose or absorbed directly by the body, were regarded as important healing agents in antiquity. The ancient custom of applying perfumes to the head and chest, consequently, was not simply an aesthetic practice but also a means of promoting well-being. 'The best recipe for health', writes the poet Alexis, 'is to apply sweet scents unto the brain.'[142] Anointing the head with perfume when drinking wine, for example, was believed to counteract the intoxicating effect produced by alcoholic fumes rising to one's head. Anointing the breast with perfume, in turn, was thought beneficial to the heart, which was held to be 'soothed with fragrant smells'.[143]

Garlands worn around the head and the breast also had the effect of supplying healthful odours to the body. Philonides, the physician, in his tract on the medicinal values of perfumes and garlands, thus recommended rose garlands to relieve head-aches and to cool the body. Myrtle garlands he held to act as a stimulant and to counteract drunkenness, while he warned against the stupefying effects of wreaths of white lilies.[144]

In the case of wounds, perfumes would be applied directly to the injury. A lotion of wine and myrrh was prescribed for burns, while megalium, the famous creation of the Roman perfumer Megallus, was thought to relieve the inflammation caused by an injury. These perfumes may indeed have promoted healing by acting as germicides.[145] In any case, they at least relieved the sufferer and his or her attendants of the foul odour emitted by festering wounds. An example of just how unbearable such an odour could be is given to us by Homer, who tells of a soldier left stranded on an island by his companions due to the stench of his wounded foot.[146] The ancients may well have thought that by counteracting the putrid odours of bodily decay with fragrance, they counteracted the decay itself.

Aromatic plants in general provided the ancients with a wide variety of curative scents. Pliny mentions a number of such aro-matic remedies in his *Natural History*. Rue in vinegar was given to comatose patients to smell as a kind of smelling salts. Epileptics were treated with the scent of thyme. The smell of pennyroyal was held to protect the head from cold or heat and to lessen thirst. The scent of a sprig of pennyroyal wrapped in wool was believed to help sufferers from recurrent fevers, while the odour of pennyroyal seeds was employed for cases of speech loss. Mint scent was thought to refresh the spirit. It was also commonly used to ease stomach-aches, as is indicated by the following epi-gram, making fun of a miser who prefers to cure his ills with the scent of money: 'Crito the miser, when he has a pain in his stomach refreshes himself by smelling not mint, but a penny piece.'[147] The smell of the *carum capticum* plant was said to help women conceive, while the smell of anise ensured an easier childbirth. Anise was also thought to relieve sleeplessness and hiccoughs through its odour and, when boiled with celery, sneez-ing. Fumigation with bay leaves, in turn, was considered to ward off the contaminating odours of disease.[148]

Healing scents were believed to emanate from certain foods as well. The fragrance of apples, much appreciated by the ancients, was held to reduce the effects of poison, while the odour of boiling cabbage was thought to soothe headaches.[149] A pupil of Epicurus is said to have denounced the cooks of his day for not knowing the medicinal value of the odours of the foods they prepared: 'How ignorant the present race of cooks are, when thus you find them ignorant of the smell of all the varied dishes which they dress.'[150] Just as there was not the same divide between perfume and food in the ancient world as there is in the modern, therefore, there were not the same divides between perfume and medicine, or food and medicine.

THE ODOUR OF DEATH

The ancients were highly concerned with the odours of death and the afterlife. This concern was perhaps most strongly expressed in the funeral practices of the ancient Egyptians. The foulness of a dead and decaying body is expressed, for example, in the following Egyptian utterance addressed to a corpse: 'How bad is your smell! How offensive is your smell! How great is your smell!'[151] Embalming, mummifying and censing the corpse were means of preventing this offensive process of decay and replacing the foul odour of death with the sweet scent of immortality.

Incense was thought by the Egyptians to provide the deceased with a scent similar to that of the gods, who were, in fact, believed to sweat incense. In one inscription a deceased king proclaims: 'My sweat is the sweat of Horus, my odour is the odour of Horus.'[152] Thus sweetened, the deceased could enter into an olfactory dialogue with the gods and request entry into their company:

Your perfume comes to me, you gods;
May my perfume come to you, you gods;
May I be with you, you gods;
May you be with me, you gods.[153]

It was believed that the deceased could actually use the incense smoke to climb up to the gods, as expressed in the following utterance: 'A stairway to the sky is set up for me that I may ascend on it to the sky, and I ascend on the smoke of the great censing.'[154] Incense therefore both made the deceased acceptable to the gods and provided the means of reaching their domain.

The ancient Greeks and Romans were familiar with the Egyptian custom of embalming, but considered it a foreign practice and only occasionally made use of it themselves. In Greece and

Rome corpses were washed and anointed with perfume. The couch on which the body was laid was sometimes strewn with flowers and, in Rome, a branch of cypress on the front door marked the house of the deceased. Incense would be burnt in the house and along the funeral procession to propitiate the gods and to ward off the ill odour of death. Nero, for example, is said to have burnt more incense than Arabia could produce in a year at his wife Poppaea's funeral. When the funeral was of someone important, flowers were also sometimes scattered along the route of the procession.[155]

After being carried in procession through the streets, the dead were either burnt or, particularly after the first century, buried. In the former case, the funeral pyre would be made of fragrant woods to which scents were added. Martial describes the olfactory stages of cremation in the following epigram.

> While the lightly-heaped pyre was being laid with papyrus for the flame, while his weeping wife was buying myrrh and casia, when now the grave, when now the bier, when now the anointer was ready, Numa wrote me down as his heir, and – got well![156]

The mythological model for cremation was the immolation of the phoenix. This legendary bird of perfumes was supposed to cremate itself on a pyre of spices when dying and then be reborn in the ashes. Similarly, the practice of cremation could be described as a process of purification, whereby the dead were released from their material bodies and transformed into pure essence, ready for their new ethereal existence in the afterlife.

When the pyre died down, the remaining flames were put out with wine and the bones of the deceased gathered. These bones would then be washed with wine, perfumed with ointment and stored in a funerary urn. When the body was buried, rather than burnt, perfumes would be sprinkled into the tomb. On regular occasions thereafter, as well, perfumes, along with food and drink, would be offered to the deceased at their grave sites.[157]

Such costly funerary customs were only for the wealthy of antiquity. Martial describes a runaway slave who makes his living stealing scents from funerals: 'The unguents and casia, and myrrh that smells of funerals, and the frankincense half-burned snatched from the midst of the pyre, and the cinnamon you have snatched from the bier of death – these, rascally Zoilus, surrender out of your foul pocket.'[158] In the first-century *Satires* of Persius, a man worries that when he dies, his heir, having received less property than expected, will neglect to perfume his bones.[159] Thus, the olfactory divide between rich and poor continued even after death.

There were various reasons for the funerary use of perfumes. One was to mask the odour of the corpse, which was considered not only unpleasant, but harmful to the living.[160] Another was to render the gods favourable to the deceased and his surviving family. A third was to provide the deceased with sweet scents, for the dead were believed to enjoy perfumes as much as or more than the living. The Elysian fields in which the virtuous dead were said to reside by the Greeks, for instance, were characterized by their sweet scents.[161]

It was the opinion of some that the afterworld consisted mainly of smoky exhalations and that the souls of the dead themselves were simply breaths of air.[162] Lucretius, for example, writes that at death 'the breath of life is driven without ... scattering abroad like smoke'.[163] Thus in Rome the closest family member would endeavour to capture the last breath of a dying relative in his or her own mouth, so as to retain, in some part, the person's soul.[164]

Lucretius also attributes the foul stench of the corpse to the loss of the vital breath, thereby setting up a dualism between the foulness of death and the fragrance of life. This association of breath with life and with the soul indicates that the importance placed by the ancients on having a fragrant breath was not simply a matter of aesthetics. To have a fragrant breath in antiquity was to exhale the sweetness of life and to attest to the purity of one's soul. Something of this sense is expressed in Martial's lamentation for his dead slave girl, Erotion:

> A maid ... whose breath was fragrant as a Paestan bed of roses, as the new honey of Attic combs, as a lump of amber snatched from the hand ... on a pyre yet new Erotion lies, whom the bitter decree of the most evil Fates carried off ere her sixth winter was full.[165]

The perfumes of the ancient funeral would provide a substitute for the fragrant breath of life, as well as a symbol of its departure.

AROMATICS AND THE GODS

According to long-standing classical tradition, the ancient gods delighted in aromatics and were aromatic themselves. Zeus is described by Homer as wreathed in a fragrant cloud.[166] Of the goddess Demeter he says that 'her odoriferous garments diffused a delectable perfume'.[167] The most fragrant of the Greek and Roman gods was the deity of love, Aphrodite, or Venus. Homer writes of Aphrodite visiting her fragrant temple in Cyprus to be anointed with ambrosia by her attendants, the Graces. When she goes, she leaves all of Cyprus sweet-smelling behind her.[168] Virgil says of Venus that 'the ambrosial locks of her hair were fragrant with heavenly odour,'[169] while an anonymous poet describes her as dressed in robes 'perfumed with the rich treasures of the revolving seasons'.[170]

The gods of antiquity played the game of scent and seduction with the same relish as mortals. Venus was said to have once given a ferryman a perfume which made him irresistible to women. While this gift provided the ferryman with a great deal of immediate pleasure, it soon proved fatal to him, for he was killed by a jealous husband who surprised him in an act of adultery.[171] Hades, god of the underworld, had the narcissus flower created to seduce Persephone by its scent, described by Homer as 'so sweet that all heaven and earth laughed with pleasure'. When Persephone reached down to pick the fragrant flower, Hades carried her down into his subterranean kingdom.[172] Even Zeus, king of the gods, was susceptible to a sweet scent. Thus, when the goddess Hera wished to seduce him, 'with ambrosia first did she cleanse from her lovely body every stain, and anointed it richly with oil, ambrosial, soft, and of rich fragrance'.[173]

Mount Olympus, the dwelling of the gods in Greek religion, was itself deemed to be a place of fragrance.[174] The earth, by contrast, was conceptualized as a place of decay and corruption. This cosmic order is comically reversed by Aristophanes in his play *Peace*. In this play the character Trygaeus uses a giant dung-beetle to take him up to a heaven befouled by the god War. While flying upwards, Trygaeus beseeches the people on the ground not to make any foul smells and to cover their excrement with perfumes, so that his dung-loving steed will not turn around and head back down to earth.[175]

Much of the ancient gods' sweet scent was due to ambrosia

and nectar, which they not only used as ointment, but also ate. Ambrosia and nectar were pure, ideal foods, fit for the sustenance of immortals. The gods, however, were also thought to enjoy feeding on the scents of burnt animal offerings and were amply provided with such by their worshippers. A wide variety of animals – from birds to oxen – were sacrificed and burnt as religious offerings in the temples of the ancient world. Such offerings might be made on a large scale on the occasion of a religious festivity, or on a small scale, by an individual desirous of a divine boon.

The supposed appetite of gods for the scents of burnt offerings was the subject of a number of ancient jests. In Plautus's *Pseudolus* a cook proclaims 'when all the pots are boiling, I take the lids off, and the odour goes flying to the sky . . . Jupiter dines on that odour every day.'[176] In Aristophanes' *The Birds* it is suggested to birds that they make the gods pay tribute for the sacrificial smoke they feed on, as to reach the divine realm the smoke first has to pass through the avian domain – the air.[177]

Along with the odour of sacrificed animals, the gods were customarily offered the fragrant smoke of incense. Incense could also constitute the whole of a sacred offering. This was particularly the case among adherents of the Pythagorean cult, which originated in Greece in the sixth century BC. Pythagoreans held that killing animals, 'who share with us a right to live and who possess a soul', was murder, and therefore emphasized the importance of incense as a bloodless offering to the gods.[178] Thus we hear, for instance, of a Pythagorean victor in the Olympic games who, unable to make the customary sacrifice of an ox to celebrate his win, instead offered up an image of an ox made of myrrh and frankincense.[179]

While incense was a standard religious offering, the gods were held to be fond of sweet scents of all sorts. Sappho writes, for example, that 'offerings of flowers are pleasing to the gods, who hate all those who come before them with uncrowned heads'.[180] Worshippers, therefore, adorned themselves and the statues of deities with garlands of fragrant flowers. Sacred statuary would also be anointed with perfume. This practice may strike us moderns, unaccustomed to perfume our images, as odd. Also contrary to modern habits was the ancient tendency to offer to the gods the same perfumes as were employed for personal use. In antiquity, however, a distinction was often not drawn between sacred and secular scents, and what humans enjoyed was presumed to be appreciated by the deities.

The addition of an olfactory dimension to sacred images and shrines was appropriate not only as an offering, but as a symbol of divine presence, for fragrance was the characteristic sign of the presence of a deity in antiquity. Thus a character in one of Euripedes' plays can sense the invisible presence of Artemis by her odour.[181] Ovid writes, in turn, that when Bacchus approached, 'the air was full of the sweet scent of saffron and of myrrh'.[182]

Experiencing the scent of the divine could have a profound effect on human beings. It is likely that divine inspiration, believed to stimulate god-like creative endeavours among humans, for example, was originally identified with the odour of divinity emitted by the gods.[183] More importantly, the odour of the gods was the odour of immortality. Ambrosia and nectar, in particular, are described in ancient literature as life-giving essences. Thus in the *Iliad*, the goddess Thetis anoints the nostrils of a fallen warrior with ambrosia to prevent his body from decaying.[184] Ovid says that the slain Aeneas was made a god when Venus touched his lifeless body with ambrosia and nectar.[185] Although the breath of life is gone, therefore, human bodies could be kept from corruption and even revived by divine fragrance.

An interesting variation on this theme is given by Ovid in *Metamorphoses*. He tells the story of how a king's daughter is seduced by the Sun and then killed and buried by her angry father. When the Sun finds the dead girl, he tries unsuccessfully to revive her with his warm rays. He then promises her that she will still be able to travel up to the sky where he resides and sprinkles her body with fragrant nectar. 'Straightway the body, soaked with the celestial nectar, melted away and filled the earth around with its sweet fragrance. Then did a shrub of frankincense, with deep-driven roots, rise slowly through the soil.'[186] Here the divine essence revives life, but in the form of a fragrant plant, which will return the favour by offering up its odour to the gods as incense.

In fact, many of the fragrant plants known to the ancients were attributed similarly legendary origins. Myrrh, for example, was said to originally have been a woman who fell in love with her father.[187] Similarly, mint was once the cast-off lover of Hades, while the laurel bush was born of a nymph beloved by Apollo.

Significantly, these plants all come into being as a result of love being thwarted in one way or another. The fragrance they exhale, consequently, might be thought of as a kind of yearning for an unfulfilled union. When the ancients offered up these aromatics to their deities, therefore, they were offering up not only pleasant scents, but whole mythological histories and an implicit desire for union with the divine.

METAPHYSICAL SCENTS

The philosophers of antiquity put forward a variety of theories and opinions concerning odour. In the fourth century BC, Plato wrote that odours partook of a 'half-formed' nature, being thinner than water and coarser than air. This ambiguous nature made odours dificult to name or classify. Plato's pupil, Aristotle, similarly remarked on the difficulty of defining odours, as opposed, for example, to colours. He concluded that the reason for this was that smells are not easily differentiated from each other.[188]

The primary olfactory differentiation made by ancient philosophers, in fact, was between pleasant and unpleasant odours. Lucretius (96–55 BC), who sought to explain the reasons for this differentiation, theorized that pleasant smells, and indeed all pleasant sensations, were composed of smooth particles, and unpleasant smells and sensations of hooked particles: 'For every shape, which ever charms the senses, has not been brought to being without some smoothness in the first-beginnings; but, on the other hand, every shape which is harsh and offensive has not been formed without some roughness of substance.'[189] Therefore tactility, according to Lucretius, underlay all sense impressions.

In the second century AD, the Greek physician Galen claimed that it was not the nose which perceived smell, but the brain. Proof of this, he held, lay in the way in which different odours were known to affect the brain. Galen characterized odours themselves as hot, cold, dry or wet.[190]

In general, the Greeks and Romans believed, in accordance with humoral theory, that the qualities of hot, cold, dry, and wet constituted the basic sensory building blocks of the cosmos. According to this system, sweet, spicy smells were associated with the characteristics of hot and dry, and rotten smells with those of cold and wet. Thus, for example, it seemed only fitting to the ancients that hot, dry lands, such as Arabia, should be the source of fine aromatics, and that the cold, wet sea should be a source of foul odours.[191] As mentioned above, the hot sun itself was associated with fragrance and the cold moon with foulness. Consequently, pleasant and unpleasant scents were not thought of simply as different aesthetic sensations in antiquity, but rather as part of a whole cosmic order.

On a metaphorical plane, olfactory and gustatory terms were often used to convey ideas of knowledge and wisdom in the classical world. The Latin word *sagax* (from which our sagacious comes), for instance, means having a keen sense of smell, and also intelligent, clever. The Latin *sapientia*, sapience, in turn, means both flavour and wisdom. Some of the colloquial associations made between intelligence and sensations of smell and taste are brought out by Plautus in the following passage from his play *Pseudolus*:

PSEUDOLUS: But about that slave who's just come from Carytus, is he pretty sharp?

CHARINUS [*holding his nose with a meaning wink*]: Well, he's pretty sharp under the armpits.

PSEUDOLUS: The fellow ought to wear long sleeves. How's his wit: pretty pungent?

CHARINUS: Oh, yes, sharp as vinegar.

PSEUDOLUS: Well, if he has to ladle out something sweet, has he any of that on tap?

CHARINUS: What a question! He's got spiced wine, raisin wine, cherry brandy, honey syrup, honey of every sort. Why he once thought of setting up a bazaar in his head.[192]

Martial, in turn, brings out the metaphorical relation between the faculty of discrimination and the faculty of smell in an epigram in which he describes an overly zealous critic as being all nose: 'Tongilianus has a nose: I know, I don't deny it. But now Tongilianus has nothing but a nose.'[193]

Indeed, the mind and the soul or life force could themselves be conceived of as 'essences' by the ancients. In Aristophanes' *The Clouds*, for example, Socrates says: 'I have to suspend my brain and mingle the subtle essence of my mind with this air, which is of like nature, in order clearly to penetrate the things of heaven.'[194] Lucretius, in turn, writes that the soul is part of the body in the same way that scent is part of a lump of frankincense.[195] The acts of emitting and inhaling odour, consequently, were not simply thought of as sensory processes, but as models for the expression and attainment of knowledge and life.

Whatever metaphysical meaning might be given to scent, when it came to the use of perfumes, ancient thinkers were sharply divided. There were those who associated perfumes with frivolity and dissipation, and those who held pleasant scents to be uplifting and invigorating. Athenaeus, in his *Banquet of the Learned*, provides a summary of the arguments of both sides. When the guests at his imaginary banquet are anointed with perfumes one of them, named Cynculus, cries out in disgust: 'Will not some one come with a sponge and wipe my face, which is thus polluted with a lot of dirt?'[196] In support of his contempt for perfume he cites a number of authorities including Socrates, who asserted that the only scent appropriate for men was that arising from the oil used for gymnastics, and Solon, who included among his laws an injunction against selling perfumes. To cap off his argument Cynculus brings up one of Sophocles' plays in which sensuality and intellect are represented respectively by Venus anointed with perfume and Minerva anointed with gymnastic oil. To this, however, another guest, Masurius, quickly retorts: 'But my most excellent friend, are you not aware that it is in our brain that our senses are soothed, and indeed reinvigorated, by sweet smells?'[197] Masurius' argument is that sensuality and intellect are interdependent, rather than opposed, and that fragrance has a positive role to play in promoting the well-being of both.

As a last note on this philosophical debate over the merits of perfume, it's interesting to learn that one of Socrates' most devoted philosophy students, Aeschines, apparently went on to take up the trade of perfumer. Of this event the rhetorician Lysias wrote ironically: 'A fine end to the happiness of this philosopher was the trade of perfumer, and admirably harmonizing with the philosophy of Socrates, a man who utterly rejected the use of all perfumes and unguents!'[198] It may be, however, that Aeschines, like the fictional Masurius, found that sensuality and intellect were not irreconcilably opposed, and that enjoying good scents, at times, made good sense.

Chapter 2

Following the scent
From the Middle Ages to modernity

With the rise of Christianity in the fourth century, the use of perfume began to fall into disfavour in the Roman Empire. Incense was condemned as part of the trappings of idolatry – 'food for demons' – the Church Father Origen called it.[1] The early Christian antipathy to incense is hardly surprising considering the number of Christians who were executed for refusing to burn incense before the image of the emperor – the standard test of imperial loyalty. Personal use of perfumes, in turn, was considered a frivolous luxury tending to debauchery by Church leaders. 'Attention to sweet scents is a bait which draws us into sensual lusts,' warned Clement of Alexandria.[2] Denial of the senses was to be the rule in all things Christian. Indeed, in their reaction against 'pagan sensuality', many Christians even ceased washing themselves and were proud to reek of 'honest' dirt and sweat – the scents allotted to the human body by its Maker.[3]

The perfumed high-life of the Roman elite, denounced as decadent by Christian and other ancient moralists, was dealt a final blow when invading Germanic tribes succeeded in dismantling the Empire in the fifth century. The 'barbarians', accustomed to a rough and ready life, had no patience for such upper-class Roman niceties as perfumed clothes and scented baths. According to the Romans, in fact, the clothes and bodies of the invaders gave off a nauseating odour. This might have been due in part to the custom which prevailed among some of the tribes of using rancid butter as a hair ointment.[4] However, this concern with 'foreign stench' is also telling of Roman fears of cultural corruption caused by the outsiders.

While much of the art and artifice of scent disappeared with the fall of the Empire, perfumes were too embedded in the ancient way of life and thought to be completely cast aside. What happened instead was that Christianity gradually incorporated and sublimated many traditional olfactory practices and beliefs. Thus by the sixth century, incense, as a symbol of prayer, had become an acceptable part of Christian ritual.[5] Fragrant flowers and odours, in turn, figured in many Christian legends, serving as symbols of virtue or miraculous signs of grace. As Christianity was adopted by the nascent European nations in the succeeding centuries, such symbolism merged with traditional folk beliefs concerning odour to create a potent new blend of scent and cosmology. The changing characteristics of this blend can be traced through the history of the West from the Middle Ages to the present century.

THE ODOUR OF SANCTITY

The most potent olfactory concept to arise from the new Christian world order was that of the 'odour of sanctity'.[6] As we saw in the previous chapter, the classical deities were frequently deemed to make their presence known through fragrance and further-more to confer aspects of their divinity on mortals through the gift of fragrant ambrosia. A mystical fragrance, similarly, was thought by Christians to signal the presence of the Holy Spirit. When manifested by an individual, this fragrance was not only a sign of divine favour, as in classical religion, however, but also a mark of the individual's exemplary holiness. Whereas the ambrosia of the classical cults had been closely linked with sensual fulfilment, the Christian odour of sanctity was clearly a sign of spiritual rectitude.

In early Christian tradition all priests were thought to emit a sweet odour in literal accordance with St Paul's statement that 'we are the aroma of Christ to God among those who are being saved' (2 Cor. 2: 15). This belief was probably reinforced by the fragrance of the rose garlands priests wore on feast days and the incense with which priests were often enveloped. The odour of sanctity, however, came to be particularly associated with persons of exceptional holiness. One early example occurs in the hagiography of the fifth-century monk Simeon Stylites. The monasticism of this period was characterized by acts of extreme asceticism – such as standing on one leg for as long as possible or staring at the sun. The form of renunciation practised by

Simeon was the rather popular one of living on top of a pillar – closer to heaven and away from worldly temptations. When Simeon was ill with fever an incomparably sweet fragrance is said to have settled around his pillar, growing in intensity until he died some days later. For his fellow believers, who were well grounded in the association of the divine with supernatural fragrance, this was an incontrovertible sign of grace.[7]

Such accounts abound in the saint-lore of the West. The thirteenth-century Blessed Herman of Steinfeld, for example, is said to have exhaled such fragrant odours that he seemed to be in a garden of delightful flowers. The seventeenth-century Venerable Benedicta of Notre-Dame-du-Laus was likewise greatly distinguished for her odours. Benedicta's body and clothes were said to be fragrant with divine perfume. This perfume scented everything she touched, growing particularly intense when the nun was in a state of ecstasy.[8]

While an odour of sanctity often appeared during the lifetime of a holy person, as in the above cases, its manifestion was particularly common on the death of such a person. When St Patrick died, for example, a sweet aroma filled the whole room where he lay. When St Hubert of Brittany died, all of Brittany was reputedly suffused with fragrance! This miraculous odour was often found to persist in the saint's body long after his or her death. The body of St Isidore, for example, showed no signs of decay and emitted a ravishing odour when it was disinterred forty years after the saint's death in the twelfth century, and then again four hundred and fifty years after his death when it was exhumed once again to be placed in a more splendid tomb.[9]

The fragrance exuded by the bodies of saints offered a striking contrast with the customary putridity of corpses, especially in an age when most people were all too familiar with the reek of death. The bodies of the well-to-do were sometimes buried with spices and herbs, but this could only supply a temporary antidote to bodily corruption. In order to forestall any suspicions that the odour of sanctity might be due to such burial practices, reports of its occurrence emphasize that no spices, ointments or balms had been used to treat the saint's body. The odour of sanctity demonstrated the power of God to place mortals outside the seemingly universal decay of death.

At the same time, the odour of sanctity stood in opposition to the stench of moral corruption. 'Some men are good smelling and some are stinking to God', asserted the fourteenth-century theologian John Wycliffe.[10] The ultimate evil odour, of course, was emitted by the Devil, with his suffocating reek of sulphur. None the less, all sins were reputed to emit a greater or lesser degree of malodour.

The concept of the odour of sanctity extended beyond the sphere of religion to that of politics; not surprisingly, given the intimate alliance between Church and State in premodern Europe. Kings, for example, were thought to derive their authority from God, as a sign of which they were anointed with holy oil. Thus Shakespeare writes in *Richard II* that

> Not all the water in the rough rude sea
> Can wash the balm from an anointed king,
> The breath of worldly men cannot depose
> The deputy elected of the Lord.[11]

The mundane breath of ordinary mortals, the playwright states, cannot avail against the sacred essence of the divinely chosen ruler. In fact, if used against the state, mundane breath would soon take on a rather offensive stink, for treachery had its stench just as did sin. The burning of traitors, together with that of witches and heretics, had the purpose not only of destroying their bodies, but also of making the malodour of their crimes known to the populace through the reek of their burning flesh. Indeed, in England, convicted traitors had their entrails removed and burnt before them as part of the execution process, so that they, along with the attending crowd, would know their vile essence.[12]

As we have seen, the moral associations of fragrance and foulness were taken quite literally in the premodern West. The inhabitants of this period consequently lived in a world that not only abounded in potent odours of all sorts, but also abounded in potent olfactory meanings, a world in which a whiff of fragrance could signify divine grace, while a sulphurous reek hinted at eternal damnation.

THE STENCH OF THE CITY

European cities were often filthy places in earlier times. Streets served as conduits for refuse of all sorts – food remains, human and animal waste, blood and entrails of slaughtered animals, and dead cats and dogs – to name some. Even the blood let from patients by barber-surgeons would be cast into the street as often as not. Most streets were made of dirt, which would mingle with waste products to produce a sticky and malodorous muck. Refuse commonly ended up in nearby rivers, either dumped there or washed down by rain, as described in the following eighteenth-century poem by Jonathan Swift:

> Now from all parts the swelling kennels [gutters] flow
> And bear their trophies with them as they go:
> Filth of all hue and odours seem to tell
> What street they sailed from, by their sight and smell.[13]

Not surprisingly, many of the watercourses of European cities were little more than open sewers. In London, the Fleet River was particularly notorious for its stench. This was so strong that the White Friar monks nearby asserted that it overpowered all the frankincense burnt at their altars and even caused many deaths among them.[14] The seventeenth-century poet Ben Jonson commemorated the malodour of the Fleet in his poem 'The Famous Voyage', which depicts two companions taking a boat trip along the river and encountering a succession of foul refuse – from animal carcasses to human waste dropped directly into the water from toilets overhead:

> ... How dare
> Your daintie nostrills (in so hot a season,
> When every clerke eats artichokes, and peason,
> Laxative lettuce, and such windie meate)
> Tempt such a passage? when each privies seat
> Is fill'd with buttock?[15]

As the population of Europe's cities increased, the problem, of course, worsened. In 1827 the chairman of the Parisian health council noted that Paris had become an olfactory landmark of a very unpleasant sort due to the abundance of rubbish dumps around the city.

> The approaches to the capital are already and *from all sides* heralded by the putrid vapors breathed there ... Soon the sense of smell gives notice that you are approaching the first city in the world, before your eyes could see the tips of its monuments.[16]

Coleridge, travelling in Germany in the late eighteenth century, had this to say about the 'eau de Cologne'.

> In Köln, a town of monks and bones,
> And pavements fang'd with murderous stones
> And rags, and hags, and hideous wenches,
> I counted two and seventy stenches,
> All well defined, and several stinks!
> Ye nymphs that reign o'er sewers and sinks
> The river Rhine, it is well known,
> Doth wash your city of Cologne:
> But tell me, Nymphs! What power divine
> Shall henceforth wash the river Rhine?[17]

While the amount of organic waste had increased immensely with the rise in population, however, it was no longer the only polluter. With the Industrial Revolution in the eighteenth century had come the added burden of industrial waste. Furthermore, the extensive use of coal fires meant that major industrial cities and towns were soon enveloped in smoky and sooty fogs. The 1837 work *London As It Is* gives a typical description of factories as 'vomiting forth ... dense volumes of black suffocating smoke, filling all the adjoining streets with stifling fumes.'[18]

The question must be asked why Europeans tolerated living in such conditions. Did they, despite all the traditional associations of foul odour with evil, simply not mind the smell? This was the view put forth by the French writer Louis-Sébastien Mercier in the eighteenth century with regard to Paris:

> If I am asked how anyone can stay in this filthy haunt ... amid an air poisoned by a thousand putrid vapors, among butchers' shops, cemeteries, hospitals, drains, streams of urine, heaps of excrement, dyers', tanners', curriers' stalls; in the midst of continual smoke from that unbelievable quantity of wood, and the vapor from all that coal ... I would reply that familiarity accustoms the Parisians to humid fogs, maleficent vapors, and foul-smelling ooze.'[19]

It is true that odours which are constantly with one recede into the background of consciousness (for which reason we usually don't notice our own odour). Yet there are enough complaints about 'noisome stenches', and enough municipal regulations designed to reduce them, on record from the Middle Ages on, to demonstrate that the inhabitants of those times were not oblivious to malodour.

Foul odours were disliked therefore, but, on the whole, tolerated. There are several reasons for this. Until the Industrial Revolution bad smells were generated primarily by organic waste and, while unpleasant, they tended to be accepted as a natural part of the cycle of life. As countryfolk knew, odoriferous manure made for plentiful harvests. As for the industrial waste which came later, ordinary people had no control over it and factory owners no wish to invest in reforms.

The municipal regulations designed to control the disposal of waste were often ineffective because they placed too much of the burden for its disposal on individual householders. While the malodour of waste in the streets and rivers was unpleasant, most city dwellers preferred that unpleasantness to the trouble of having to continually cart off their waste. Furthermore, particularly after the Industrial Revolution, many people simply spent too much of the day working to have time left over for practices of cleanliness.

Local governments, however, were not ready to spend the time and money needed to create a more efficient, centralized, system of waste disposal. At the most, street cleaners would be hired, but these were not numerous enough to keep the cities free of waste. Municipal services were generally very few in Europe prior to the twentieth century. Every once in a while, especially during plague outbreaks when the fear of infection by 'corrupt air' ran high, serious attempts would be made to keep streets clean. None the less, even in this there was not complete agreement, for some believed that waste odours helped ward off disease, as well as 'tempering' air which might otherwise be excessively pure for human consumption. Even industrial smoke had its supporters. The author of *London as It Is* notes: 'Many persons think that the smoke is beneficial rather than prejudicial to health in London, on the idea, probably, that it covers all other offensive fumes and odours.'[20]

As for how people who associated foul odours with sin could live amidst so many of them, the evident answer is that they also lived among sins of all sorts, and that the foulness of their environment only served to remind them of their sinful condition. In Heaven there would be fragrance, on Earth there was corruption and stench.

At the same time, malodours were not necessarily considered to be evil. A sixteenth-century allegory, for example, tells the story of an angel who has no objection to the smell of an honest dung-collector's laden cart, but who stops up his nose at a perfumed courtesan.[21] Even when foul smells *were* associated with ungodliness, the attitude of many towards them probably echoed the sentiments expressed in the following sixteenth-century verse, in which a priest responds to a devil who mocks him for praying while on the privy:

> Ech take his due, and me thou canst not hurt,
> To God my pray'r I meant, to thee the durt.
> Pure pray'r ascends to Him that high doth sit,
> Down fals the filth, for fiends of hel more fit.[22]

PLAGUES AND POMANDERS

Plagues were an unavoidable part of life in medieval Europe, occurring every few years from the fourteenth century until the end of the seventeenth century. Science has now determined that these pandemics were spread primarily by rat fleas carrying plague germs. At the time, however, the humble flea was not suspected and other, more dramatic, agents were put forward as responsible for the deadly disease. Many thinkers held that plagues were caused by astrological influences: a change in the alignment of the planets; or an increase in strength of the rays of the sun and Syrius, the dog star.

This latter reason sought to explain why the plague was most active during the hot 'dog-days' of summer, when Syrius rises and sets with the sun. An anonymous seventeenth-century poet expressed this belief in verse form:

> How falsely doe old Poets speake when they
> The Sun the God of Physick [medicine] call
> When as we see that by his burning ray
> He cures not any, but doth murther all ...
> The dogstar strait put forth his head
> Yet could not long look on
> But blushing with a more than nat'rall red
> Retired, as if ashamed of what he had done.
> But yet thou angry Star, we learne a way
> By killing dogs thy cruelty to repay.[23]

This last line refers to the slaughter of dogs which took place in many plague towns, for these animals were commonly believed to transmit the disease.

By far the most widely accepted cause of the plague, however, was foul odour caused by putrefaction. In the words of a medieval alchemical poem:

> And when evyl substance shal putrefie
> Horrible odour are gendered therbye
> As of dragons, and men that long dead be,
> Their stinke may cause great mortalitie.[24]

Put succinctly, putridity engendered putridity, with smell constituting the primary agent of contagion. For those not content to leave it at that, this theory was easily combined with grander designs. Martin Luther, seeking a theological basis for the plague, thought that evil spirits 'poisoned the air or otherwise infected the poor people by their breath and injected the mortal poison into their bodies'.[25] Those with a scientific turn of mind held that it was the sun and the planets which fouled the air with invisible poisons. Jupiter, it was said, being a warm and humid planet, drew up putrid vapours from the earth, which the hot and dry Mars then ignited, returning them to earth as a pestilential gas. Others argued that the corrupt air came not from the planets, but from within the earth itself and was released into the atmosphere through earthquakes.[26]

This belief in pathogenic odours as plague carriers was strengthened by the fact that victims of the plague themselves emitted a strong smell. One writer observed that 'all the matter which exuded from their bodies let off an unbearable stench... so foetid as to be overpowering'.[27] Whether or not belching earthquakes or planetary gases engendered the plague, it was clear that it *could* be caught through contact with a person who already had it, and smell, as that characteristic of the disease which most evidently was emitted by the ill and internalized by the well, appeared the logical medium of contagion.

The overwhelming putridity of plague victims seemed to many God-fearing people to be none other than the reek of sin made manifest. How many times had church leaders railed against the unholy stench of the vice-ridden cities of Europe? Here then was the just consequence of that vice – cities full of rotting bodies. If a rotten soul produced a rotten body, however, it should, God willing, be possible to keep one's body well by purifying one's soul. One seventeenth-century plague tract thus advised cleansing the stink of worldly love with 'the spoone of faithfull Prayers', and warming one's heart with repentance in order to 'sweate out all the poyson of covetousnesse, pride, whoredome, idolatrie, usury, swearing, lying, and such like'.[28]

Such religious remedies aside, measures against the plague were directed in large part at controlling and combatting corrupt air. Virtually any pungent odour was thought to be good for this purpose. Municipal authorities had bonfires of aromatic woods burnt in the streets to purify the atmosphere. Individuals fumigated their homes with, among other things, incense, juniper, laurel, rosemary, vinegar and gunpowder. Even burning old shoes was thought to help, while, for added olfactory protection, some families kept a goat in the house.[29]

The most cautious inhabitants enclosed themselves in their homes with a stockpile of food and refused to see, and especially *smell*, anyone until the plague had abated. Many others fled the diseased cities for the countryside only to be turned away by plague-fearing country dwellers. Those who stayed in town made sure to carry with them some olfactory prophylactic whenever they ventured outside. One of the most popular of such devices was the pomander – originally an orange stuck full of cloves and later any perforated container filled with scents and carrried on the person. Otherwise a cautious person might carry a bouquet of aromatic flowers or a handkerchief sprinkled with perfume.

Special care was taken when entering a sickroom. One London physician recommended that the sickroom have herbs at the windows, an aromatic fire burning in the fireplace, and rose-water and vinegar sprinkled on the floor, when visitors were to be received. Visitors were advised to wash themselves with rose-water before entering, to keep a piece of cinnamon or other spice in their mouths, and to carry a pomander to smell. On leaving, another wash with rose-water was in order. Physicians themselves sometimes wore a 'nose-bag' filled with herbs and spices over their noses when visiting patients.[30] The use of such odorants was not, in fact, without medicinal value, for, while inadequate for stopping the plague, the essences of many aromatic plants, such as lavender or garlic, are powerful germicides.

Aromatics were considered useful not only for preventing disease, but also for curing it. Contemporary medical theory held that the nose gave direct access to the brain. Medications inhaled through the nose, therefore, were reputed to act more directly on the brain, hence the spirit, than those swallowed. Furthermore, the spirit, or life force, was imagined to be similar in nature to odour, making smell the best means of correcting its disorders.[31]

It is to be imagined that perfumers and flower-sellers did very brisk business in times of plague. Thomas Dekker writes of the Great Plague in England that 'the price of flowers, hearbes and garlands, rose wonderfully, in so much that Rosemary which had wont to be solde for 12 pence an armefull, went now for sixe shillings a handfull'.[32] For the many who could not afford such expensive remedies, simpler, less fragrant, measures made do. A humorous poem describes one cautious Londoner who

> with a peece of tasseld well tarr'd Rope
> doth with that nose-gaye keepe himselfe in hope,

and another who

> takes off his socks from's sweating feete
> and makes them his perfume along the streete.[33]

In his semi-fictional *Journal of the Plague Year*, Defoe describes a gravedigger whose preservative against infection consisted of 'holding garlic and rue in his mouth, and smoking tobacco'. The gravedigger's wife, who worked as a nurse, protected herself by 'washing her head in vinegar ... and if the smell of any of those she waited on was more than ordinary offensive, she snuffed vinegar up her nose.'[34] Even odours of excrement and urine (provided these came from a healthy source) were considered by some to protect against infection by virtue of their potency. Thus in 1680, a German physician reported a plague practice of 'standing over a privy in the early morning, to inhale the stink on an empty stomach'.[35] Swift mocks such practices in *Gulliver's Travels*, in which his malodorous parodies of human beings, the Yahoos, treat themselves when ill with a mixture of their own dung and urine.[36]

All in all the plague years, with scores of sick and dead bodies reeking of decay, scented fires burning in streets and houses, and aromatic remedies of all sorts employed, were ones of strong olfactory stimuli. Wherever people gathered, a fantastic mélange of scents – rosemary, lavender, juniper, garlic, vinegar, cloves, tar, perspiration, and countless others – would have filled the air.

Defoe writes of a well-attended church service during a plague outbreak that 'the whole church was like a smelling-bottle; in one corner it was all perfumes; in another, aromatics, balsamics, and a variety of drugs and herbs; in another, salts and spirits.'[37]

Such olfactory imagery would have been powerful not only in itself, but in its emotional associations. The smells of home and town were the smells of an olfactory war, an immense battle against the deadly odours of corruption in which every pungent scent available was enlisted. No one was above suspicion in this war, for even people who had the appearance of health might already be contaminated and emitting noxious airs. The plague, therefore, had the effect of generating immense anxiety about the odours of others, each person desiring to remain enclosed in a private olfactory bubble, shielded by walls of rosemary or cloves. At the same time it reinforced concepts of odour as a primary force for good or ill, holding the power of life and death. In the years between plague outbreaks, when the normal routine of life was followed, olfactory phobias declined but the idea of odour as a vital force continued to hold sway.

HOME, SWEET HOME

Medieval and Renaissance homes were a combination of fragrance and foulness. Which of the two predominated in any particular case depended not only on the wealth of the inhabitants, but also on their habits of housekeeping. The Dutch humanist Erasmus, for instance, wrote of English houses in which:

> The floors are made of clay, and covered with marsh rushes, constantly piled on one another, so that the bottom layer remains sometimes for twenty years incubating spittle, vomit, the urine of dogs and men, the dregs of beer, the remains of fish, and other nameless filth. From this an exhalation rises to the heavens, which seems to me most unhealthy.[38]

The houses which his fellow countryman, Leminus, visited in England evidently changed their floor coverings more frequently, for he commented that:

> Their chambers and parlours strawed over with sweete herbes refreshed me; – their nosegays finely intermingled with sundry sorts of fragraunte floures, in their bed chambers and privi

rooms with comfortable smell cheered me up and delyghted all my senses.[39]

Ill-kept homes, depositories of waste of all sorts, could therefore be incredibly foul. A well-kept home, regularly cleaned and aired and scented with flowers, on the other hand, could be pleasantly fragrant – at least during the summer months when fresh flowers and rushes were available, and sooty fires and malodorous tallow candles not in use.

The olfactory condition of a habitation was not considered simply a matter of sensory pleasure or displeasure during this period, however, but also of physical well-being. As we have seen, odours were believed able to both communicate and prevent disease, and this characteristic of smell informed contemporary evaluations of household scents. Erasmus, for example, describes the odour arising from soiled rushes not as disgusting, but as unhealthy.

As previously noted, however, unpleasant smells were not always thought harmful. The widespread replacement of the traditional hole in the roof to draw out smoke by the more efficient chimney in the sixteenth century was resisted by many who thought a smoke-filled house good for the health. A conservative English rector noted, for instance, that before the introduction of chimneys, colds had been a rarity, 'for as the smoke in those days was supposed to be a sufficient hardening for the timbers of the house, so it was reputed to be a [good] medicine to keep the goodman and his family from the quack.'[40]

Apart from being smoky, the houses of the Middle Ages and the Renaissance were often damp and musty. In the countryside an additional olfactory note might be added by the presence of farm animals, who often shared the same roof with their owners. Ulrich von Hutten, writing in 1518, evokes the characteristic scents of a castle:

The castle was built not for pleasure but for defense, surrounded by moats and trenches, cramped within, burdened with stables for animals large and small, dark buildings for bombards and stores of pitch and sulfur ... Everywhere the disagreeable odor of powder dominates. And the dogs with their filth – what a fine smell that is![41]

Furthermore, sanitation in homes of all classes was very basic.

Rooms were furnished with chamber-pots or buckets which would be emptied into a nearby stream or cesspit, or simply into the street. A larger home would have a projecting room on the upper storey with a hole in the floor positioned over a moat, river or side alley for the disposal of waste.

Such methods, not surprisingly, often generated unpleasant odours. King Henry of England complained in 1246, for example, that his 'privy chamber' was 'situated in an undue and improper place, wherefore it smells badly'.[42] The problem was compounded by the fact that, not only privies and chamber-pots, but almost any corner of a house, from fireplaces to cellars, might on occasion be used by ill-mannered persons to relieve themselves.[43] In larger houses and palaces the situation could be serious. Versailles, for instance, for all its visual splendours, stank of urine and excrement.[44] A certain tolerance of waste odours was therefore, as noted above, a condition of the times.

The use of fragrance in the home was recommended as a means of sweetening and purifying the air. The simplest method was that mentioned by Leminus above – strewing the floors with herbs, which would release a pleasant odour when stepped on, and keeping nosegays in the different rooms. Thyme, basil, camomile and sweet flag were popular strewing herbs, while wallflowers, marjoram, and sweet william were favourites for nosegays. Such aromatic plants as lavender and rosemary were often grown indoors in pots and placed on window-sills or hung from the ceiling to refresh the atmosphere.

Perfumes might also be used to sweeten a house. Rose-water, for example, would be sprinkled around rooms from a perforated bottle known as a casting bottle. Another common method of making a room fragrant was to burn aromatic wood, such as juniper or briar, in the fireplace, or simmer scented water in a pan. Dried lavender stems, in turn, provided a pleasant and readily available incense. For the wealthy, there were professional perfumers, who travelled the country fumigating the musty rooms of manors and castles.[45]

With regard to household furnishings, wood furniture and panelling were rendered fragrant by the practice of polishing them with sweet herbs. Pillows and cushions, for their part, might be filled with dried herbs such as woodruff. In churches, the pews of the gentry were sometimes strewn with flowers, as we read in the sixteenth-century *Apius and Virginia*:

My lady's fair pew had been strewn full gay
With primroses, cowslips and violets sweet,
With mints and with marygold and marjoram meet.[46]

In the homes of the well-to-do, hangings and bedlinen were commonly perfumed, while aromatic sachets or powders kept clothes smelling sweet in their storage chests. In fact, the perfumes used for these purposes were sometimes so strong that onlookers would be left gasping for air when such a chest was unpacked.[47]

Historical accounts provide us with an intriguing record of the perfumes of royal households. In England, Queen Elizabeth I preferred her apartments to be strewn with meadowsweet. The Elizabethan gardening author John Gerard wrote of this herb: 'the leaves far excell all other strewing herbs to deck up houses, to strew in chambers, hall and banqueting houses in summertime, for the smell thereof makes the heart merry and joyful.'[48] Rosewater and sugar boiled together made the room of Edward VI smell 'as though it were full of roses',[49] while rosemary and sugar perfumed the chambers of Queen Anne. George III is said to have used a pillow filled with fragrant hops as an aid to his slumber.[50]

Much of the fragrance that came into the home – herbs, nosegays, floral waters – came from the garden outside the home. The medieval garden was traditionally walled, so that the scent of the aromatic herbs and flowers grown inside was intensified by being confined in an enclosed space. Renaissance gardens were more spacious and elaborate than their predecessors; but the emphasis on fragrance continued. In his classic gardening essay from 1625, Francis Bacon lists in detail the plants which contribute to the olfactory beauty of a garden: violets, musk roses, strawberries . . .

> But those which Perfume the Aire most delightfully, not passed by as the rest, but being Troden upon and Crushed, are Three: That is Burnet, Wilde-Time, and Water-Mints. Therefore, you are to set whole Allies of them, to have the Pleasure, when you walke or tread.[51]

Apart from its aesthetic value, the garden, with its fresh scents, was also considered an important bulwark against the corrupt odours of disease. This dual function is outlined in a couplet which describes the garden at Hampton Court in England as having:

arbours & alleys so pleasant & so dulse
the pestilent airs with flavours to repulse.[52]

It was not until the eighteenth century, when landscape gardening
came into vogue, that gardens began to lose their aromatic intimacy in favour of sweeping vistas of green lawn and trees. This
new style affected only large estates, however, while cottagers
continued to grow their lavender, rosemary and rue, together
with their cabbages and onions, in their garden plots.[53]

THE SCENTS OF DINNER

First they brought him delicate wine,
Mead in bowls of maple and pine,
All sorts of royal spices,
And gingerbread as fine as fine
And licorice and sweet cummin,
And sugar that so nice is.

The Tale of Sir Topaz, *Canterbury Tales*[54]

As in the classical period, herbs and spices were widely used to
flavour food in medieval and Renaissance Europe. While the
former were readily available to all, however, the latter were
luxury goods from the fabled East. Spices had been introduced
to medieval Europe by the crusaders, who acquired a taste for
them during their sojourns in the Holy Land, and pepper, ginger,
cloves, cinnamon and nutmeg quickly became essential ingredients of upper-class Western cookery. Due to their high price,
spices were items of prestige, suitable not only for seasoning, but
for gift-giving and bequeathing to one's heirs. At the same time,
the general belief that spices came from the Garden of Eden,
made their consumption a quasi-religious experience – breathing
in the scent of spices, one could feel that one was catching a
whiff of Paradise.

Therefore, while peasants made do with simple dishes, such as
bread, ale and pottage – vegetable stew – and simple seasoning,
such as mint, garlic and onion, medieval banquets were rich
and spicy affairs. Meats were customarily seasoned with such
condiments as pepper, cinnamon, cloves and ginger (and often
sugar as well, for the medievals had little compunction about
mixing up flavours). This heavy seasoning would have served to
disguise any decay, but, more importantly, it satisfied the medieval

craving for highly flavoured foods. Other spiced dishes included fruit stews, porridges, puddings and pies. Drinks – ales and wines – would also be scented with spices. Finally, in order to enjoy the cherished condiments in their integrity, a spice platter, containing such aromatic delicacies as pepper grains, cinnamon sticks, crystallized ginger and perfumed sugar would often be passed among the guests.[55]

In this feast of scent, flowers had a role to play as well. Rosewater was used to flavour a variety of dishes. (In the absence of cutlery, it would also be offered to the guests for washing their hands.) Violets lent a sweet fragrance to stews. Various flowers, in particular roses and carnations, gave their special scent to wines. Roses and violets were added to puddings, pies and cakes, or candied and used as garnishes. Other floral favourites of the kitchen included orange flowers, hawthorne blossoms and primroses. One sweet and spicy medieval dish, for example, consisted of ground primroses, almonds, rice flour, almond milk, honey and saffron, simmered together and sprinkled with powdered ginger.[56]

This style of cookery continued into the Renaissance. The heavy demand for spices led to an avid search for better trade routes to the East and eventually to the New World, which, in turn, offered a new range of flavours to the European palate. Potatoes, tomatoes and corn came to the Old World from the New, along with vanilla, chilli peppers, and tobacco. This last particularly excited the curiosity of Europeans, who had never *smoked* anything before. Legend has it that the first time a servant of Sir Walter Raleigh saw his master smoking a pipe, he thought he was on fire and doused him with water. Smoking soon found a niche in Western culture, however, both as a pastime and as a means of fumigation. Enveloped in a cloud of aromatic tobacco fumes, the seventeenth-century pipe-smoker felt not only comfortably relaxed, but also protected against the invasive odours of disease.[57]

Renaissance feasts were as aromatic as those of the Middle Ages. One sixteenth-century cook describes a papal banquet as follows:

There were dishes made with rose-water; on the same dish the most varied ingredients might be combined. The union of opposites was the triumph of culinary art. Before dessert, the cloth was removed; hands were washed; the table was covered with sugared eggs and syrups, which sent forth numbing fragrances. At the conclusion, bouquets were distributed.[58]

In addition to their culinary scents, such banquets would be sweetened by incense burning in a dish, or more stylishly, in the beaks of imitation peacocks. At one royal banquet in Naples attended by Charles V, peacocks and pheasants stuffed with spices were served; when carved they filled the whole room with their scent.[59]

In the seventeenth century, however, the use of spices in cookery began to decline. By this time, their pungent savours were no longer an exciting novelty and their mystical aura had worn thin. The Puritan movement contributed to this decline by denouncing spices as sensual stimulants, exciting all of the grosser passions. Serious-minded Christians, it was argued, should not indulge in such gastronomic excess, but keep to plain, unpretentious fare. The blandness of the *nouvelle cuisine* was counterbalanced, however, by stimulating new drinks – coffee, tea and chocolate (originally consumed as a beverage). All three of these were, as spices had been, exotic luxury items at first. Coffee had come to Europe from Arabia, tea from China and chocolate from Mexico. The intense demand for these savours soon led, as in the case of spices, to new trade relations and commercial empires, bringing wealth to some and slavery to others in overseas plantations.[60]

As their prices dropped with increased production in the eighteenth century, the aromatic hot drinks became popular with all social levels. Some disdained the new beverages – 'tea makes me think of hay and dung, coffee of soot and lupine-seed, and chocolate is too sweet for me,'[61] one German aristocrat complained – but they were in the minority. The drinks met with the approval of the Puritans because they provided relatively harmless substitutes for the alcoholic beverages heavily consumed by peasants and nobles alike. Coffee is so much the smell and taste of morning for us now, that it is odd to think that, before its advent, beer was the common breakfast drink in Europe. With the introduction of coffee, tea and chocolate, Europe sobered up.[62]

Let us conclude with a look at the foul underside of Western cuisine. With little means of refrigeration, the meat and fish which were sold and served in premodern Europe were at times rotten. In 1366, for example, an Englishman was convicted of selling thirty-seven putrid pigeons. His sentence was to be shackled in the pillory and censed with the malodorous smoke of the pigeons burning beneath him. Numerous recipes of the time dealt with different methods of disguising the taste of spoiled meat, such as soaking the meat in vinegar or smothering it with pungent sauces. Still, perhaps making a virtue of a necessity, many people seemed to *prefer* their meat putrid, or 'high', so that it gave off a strong odour when served at the table.[63]

The kitchen itself often reeked of the smell of refuse, spoiling food, cooks sweating from the heat of the fire, dogs turning the spit and animals being slaughtered for supper. From the methods used to accomplish this slaughter it would seem that the more an animal was tortured before it died, the tastier a dish it was believed to make. Thus, in eighteenth century cookery, for example, eels would be impaled through their eyes and skinned alive, pigs whipped to death with knotted cords and geese cleverly roasted and carved on the table while still (barely) alive. 'It is mighty pleasant to behold!!!!' enthused the propagator of this last 'recipe', a Dr William Kitchener.[64] If spices brought one a whiff of Paradise, the rites of the kitchen must have often conveyed the scent of hell.

Moving to the dining room, there were a number of potential sources of stench besides spoiled food. Slovenly servants were one such source. In an eighteenth-century satirical guide to servants, Jonathan Swift advises footmen to 'never wear Socks when you wait at Meals' because 'most Ladies like the smell of young Men's Toes'.[65] The diners themselves, whatever their social class, could also be unpleasantly odoriferous. In order to allow diners to relieve themselves on the spot, for example, chamber-pots might be kept in the dining room.[66]

The following passage from *The Arabian Nights* offers an imaginative description of Western foulness of food and person from a fastidious non-Western standpoint:

> They eat evil-smelling, putrescent things, such as rotten cheese and game which they hang up; they never wash, for, at their birth, ugly men in black garments pour water over their heads, and this ablution, accompanied by strange gestures, frees them from all obligation of washing for the rest of their lives. That they might not be tempted by water, they destroyed the [public baths] and public fountains, building in their places shops

where harlots sell a yellow liquid with foam on top, which they call drink, but which is either fermented urine or something worse.[67]

A biased portrait, certainly, but one with an odorous grain of truth in it.

THE PERFUMED BODY

When considering the odours of yesteryear it is important to keep in mind that standards of personal cleanliness were quite different in the premodern West from what they are today. Prior to the eighteenth century bathing tended to be considered more as a sensual pastime, and therefore somewhat decadent, than as a means of cleaning oneself. Thus St Francis of Assisi, for example, included dirt as an insignia of holiness. Furthermore, it was thought that water not only morally corrupted the body, but physically corrupted it as well by rendering it moist and soft – feminine – and vulnerable to unhealthy air and disease.

The dangers believed to be inherent in bathing can be seen by the precautions which were observed when the rare bath *was* taken. Francis Bacon, for instance, gave the following prescription for a bath which seems more like an elaborate perfuming than a wash with water.

> First, before bathing, rub and anoint the Body with Oyle, and Salves, that the Bath's moistening heate and virtue may penetrate into the Body, and not the liquor's watery part: then sit 2 houres in the Bath; after Bathing wrap the Body in a seare-cloth made of Masticke, Myrrh, Pomander and Saffron, for staying the perspiration or breathing of the pores, until the softening of the Body, having layne thus in seare-cloth 24 houres, bee growne solid and hard. Lastly, with an oyntment of Oyle, Salt and Saffron, the seare-cloth being taken off, anoint the Body.[68]

When Henri IV of France, on requiring the presence of one of his ministers, learned that the man was taking a bath, he insisted on putting off the meeting until the next day: 'He orders you to expect him tomorrow in your nightshirt, your leggings, your slippers and your night-cap, so that you come to no harm as a result of your recent bath.'[69]

There were some people who took baths fairly often in spite of their reputed dangers. Elizabeth I of England, for example, reportedly took a bath once a month 'whether she need it or no'.[70] For most, however, washing was restricted to the hands and perhaps face. The gentry used scented water for this purpose, as described by Shakespeare in *The Taming of the Shrew*:

Let one attend him with a silver basin
Full of rose water, and bestrew'd with flowers.[71]

Such toilet waters were preferred to soap, which, being made of tallow or whale oil and potash, was often too coarse and foul-smelling to be used on the skin. The body might be cleansed by being rubbed with a scented cloth. 'To cure the goat-like stench of armpits', writes a sixteenth-century French hygienist, 'it is useful to press and rub the skin with a compound of roses.'[72] In fact, clothes themselves were regarded as cleansing the body of dirt. Washing one's clothes therefore, served the same purpose as washing one's body, and with much greater safety. Hair, in turn, was cleaned by being rubbed with scented powders. The breath was freshened by chewing herbs such as aniseed, rinsing the mouth with cinnamon or myrrh water, or sucking on perfumed candies – 'kissing comfits'.

The importance of perfuming oneself lay in the fact that perfumes were not just thought to mask unpleasant odours, but to actually dispel them. Furthermore, fragrance was held to be therapeutic, serving to strengthen and stimulate mind and body. Aside from these practical considerations, however, Europeans took immense pleasure in perfume. This pleasure reached a height of expression in the sixteenth and seventeenth centuries when, among the wealthy, everything from letters to lapdogs was scented.

Apart from the customary floral fragrances and the imported spices, scents of animal origin – musk, civet and ambergris – were very popular during the Renaissance. Musk was extracted from a scent gland of the musk deer, native to India and China, and civet from that of the civet cat of Ethiopia and Indonesia. Ambergris is an excretory product of the sperm whale found floating on the ocean or washed up on shore and used primarily as a fixative for other scents. Its source remained a mystery until the start of the whaling industry in the eighteenth century, when whalers found lumps of it inside the whales they were cutting up.

For Westerners, these exotic substances were invested with legendary qualities. Much as the gathering of cinnamon was once thought to be fraught with risk, for example, stories were now told of the extraordinary cunning and skill required by the hunter of the musk deer. (In fact, musk deer have been hunted almost to extinction for their precious pods of musk.) Furthermore, due to their animal origin, musk and civet were believed to radiate a potent natural vitality. This led to their use as olfactory aphrodisiacs by amorous ladies and gentlemen. Thus, when a character in *Much Ado About Nothing* rubs himself with civet, it's 'as much as to say, the sweet youth's in love'.[73]

Fragrances were used singly or blended together to form compound perfumes. Rose and musk was one common combination, and a favourite of both Henry VIII of England and his daughter Elizabeth. Not only the body would be perfumed, however, but virtually everything worn on the body as well. Clothes were washed with lavender and dusted with aromatic powders. Gloves and shoes were made of perfumed leather and chosen as much for their scent as for their appearance. Ornate pomanders containing musk and spices, and pouncet boxes full of perfumed powders or snuff, were carried in the hand or worn around the waist or neck. Bracelets and necklaces were made out of beads of hardened perfume. Rings concealed grains of scent in tiny, perforated boxes. Even gemstones might be odorized with perfume. (In fact, this practice may have been thought of as simply bringing out the gem's innate essence, for one theory had it that stones were originally made of water condensed by odour.)[74] Taking into account all these different ways in which perfumes were worn on the person, in addition to natural body odours, the gentry of this period must have been odoriferous indeed.

Perfume, however, was not considered simply something to be passively bought and worn, but a means of diversion. Ladies, and sometimes gentlemen, of the court enjoyed making floral waters at their own stills and creating personal blends of fragrance. As in the days of ancient Rome, perfumes often formed part of the entertainment on social occasions. One elaborate seventeenth-century plan for a banquet, for example, had the guests throwing eggshells filled with rose-water at each other. At a banquet given in Naples in 1476, a miniature fountain spraying orange-flower water adorned the table. On an occasion when Queen Elizabeth entertained a delegation of French ambassadors, 'two cannons were shot off, the one with sweet powder, and the other with sweet water, verie odoriferous and pleasant.'[75]

This olfactory largesse was continued among the aristocracy of the eighteenth century. At the Versailles court of Louis XV, known as 'la Cour parfumée', fashion dictated that a different perfume be worn each day of the week. In order to ensure that she would never be left without a scent, the king's lover, the Marquise de Pompadour, reportedly spent a million francs creating a perfume bank.

One perfume which was catching on in the seventeen hundreds was Eau de Cologne. Created by Italian perfumers living in Cologne and composed of rosemary and citrus essences dissolved in grape spirit, Eau de Cologne had originally been a plague preventive. By the nineteenth century it was enormously popular as a perfume all over Europe. Napoleon was said to be so fond of this scent that he would splash a vial of it over his head every morning.[76]

None the less, there were many during these centuries of extensive perfume use who disapproved of perfume. Chief among these were Protestant reformers, such as the Puritans. Their views on perfume were those of the early Christians: it encouraged personal vanity and licentiousness. Moreover, perfume disguised humanity's innate state of corruption with its artificial sweetness. Bodies which are now 'so perfumed and bathed in odoriferous waters', warned a seventeenth-century pamphlet, 'must one day be throwne (like stinking carrion) into a rank & rotten grave'.[77] These views did not immediately stem the outpouring of scent, but they did have an effect on popular attitudes towards perfume in the centuries to come.

In any case, as we shall see below, fragrance trends were changing. By the late eighteenth century musk and civet were no longer in favour as perfumes. They were too strong, too animalistic, their 'excremental' odours repulsed, rather than attracted, persons with a refined sense of smell. The new perfume ideal was that of delicate floral and herbal scents: lavender, rosemary, violet, thyme, rose. There were exceptions to this rule – the Empress Josephine, for example, adored musk as much as her husband did Eau de Cologne – but on the whole the new floral ideal would dictate perfume fashion until the twentieth century.[78]

ODES TO ODOUR

The sixteenth and seventeenth centuries were a period rich in olfactory verse, or 'odes to odour'. This was partly due to classical influences, for the writers of that time were very familiar with and admiring of classical works, in which olfactory imagery abounds. It is evident, however, that the olfactory odes of those centuries are no mere reworkings of a dead and ancient theme, but vigorous expositions of a thriving symbolic system of odour.

These poetic descriptions of scent deal with any number of topics – the fragrance of nature, the stench of city life, the odours of disease and so on. John Donne's poem entitled 'Elegy: The Perfume' describes a lover who, though employing all manner of precautions to avoid discovery when visiting his beloved, finds he has been given away by his perfume.

> But oh, too common ill, I brought with me
> That, which betray'd me to my enemy:
> A loud perfume, which at my entrance cried
> Even at thy fathers nose, so were we spied ...
> I taught my silks, their whistling to forbear,
> Even my opprest shoes, dumb and spechless were,
> Only thou, bitter sweet, whom I had laid
> Next me, me traitorously hast betrayed,
> And unsuspected hast invisibly
> At once fled unto him, and stayed with me.[79]

Not the loud perfume of the lover, however, but the sweet scent of the beloved furnished one of the most popular themes for odes to odour. Restricting ourselves to English literature, Edmund Spenser offers us a typical example of such an ode in his 'Sonnet 63', in which the beauty of the beloved is expressed in terms of a sequence of floral scents:

> Comming to kiss her lyps, (such grace I found)
>   Me seemed I smelt a gardin of sweet flowres:
>   that dainty odours from them threw around
>   for damzels fit to decke their lovers bowres.
> Her lyps did smell lyke unto Gillyflowres,
>   her ruddy cheeks lyke unto Roses red:
>   her snowy browes lyke budded Bellamoures,
>   her lovely eyes lyke Pincks but newly spred.
> Her goodly bosome lyke a Strawberry bed,

her neck lyke to a bounch of Cullambynes:
her brest lyke lyllyes, ere theyre leaves be shed,
her nipples lyke yong blossom'd Jessemynes.
Such fragrant flowres doe give most odorous smell,
but her sweet odour did them all excell.[80]

Robert Herrick's poetry particularly abounds with imagery of this sort, of which his 'Upon Julia's Sweat' is an instance.

Wo'd ye oyle of Blossomes get?
Take it from my Julia's Sweat:
Oyl of Lillies, and of Spike,
From her moysture take the like:
Let her breath, or let her blow,
All rich spices thence will flow.[81]

As among the classical authors, such laudations of scent were counterbalanced by frank deprecations, for example, Herrick's 'Upon a Free Maid, with a Foule Breath'.

You say you'll kiss me, and I thank you for it:
But stinking breath, I do as hell abhor it.[82]

The most damning of these can be found in Jonathan Swift's verses of the following century. In 'The Lady's Dressing Room', he mercilessly dwells on the foul underpinnings of the outwardly elegant lady of fashion:

The Stockings, why should I expose,
Stain'd with the Marks of Stinking Toes,
Or Greasy Coifs and Pinners reeking,
Which Celia slept at least a week in?[83]

In 'Strephon and Chloe' he mocks the impossible ideal of the perpetually fragrant beloved:

No Humours gross, or frowzy Streams,
No noisome Whiffs, or sweaty Steams,
Before, behind, above, below,
Could from her taintless body flow.[84]

The apparent delight certain poets took in describing foul scents reminds us that such odours were not yet excluded from public discourse, as they would be in the nineteenth century, although an etiquette promoting such exclusion was not unknown. Poets such as Swift, in fact, were undertaking in verse precisely what etiquette guides of the period advised people not to do in actuality. A guide of 1609, for example, warns:

> it is not a refined habit, when coming across something disgusting in the street, as sometimes happens, to turn at once to one's companion and point it out to him.
>
> It is far less proper to hold out the stinking thing for the other to smell, as some are wont ... lifting the foul-smelling thing to his nostrils and saying, 'I should like to know how much that stinks.'[85]

Smell, however, is not only a matter of aesthetics or ribaldry in the poetry of this period, but also of morality. In Shakespeare's 'Sonnet 54', for instance, the scent of the rose stands for abiding truth and virtue.

> O how much more doth beautie beauteous seeme
> By that sweet ornament which truth doth give;
> The Rose looks faire, but fairer we it deeme
> For that sweet odor which doth in it live.[86]

Similarly, in George Herbert's 'Life', sweet scents are equated with moral value.

> I made a posie, while the day ran by:
> Here will I smell my remnant out, and tie
> My life within this band.
> But time did beckon to the flowers, and they
> By noon most cunningly did steal away,
> And wither'd in my hand ...
>
> Farewell, dear flowers, sweetly your time ye spent,
> Fit, while ye lived, for smell or ornament,
> And after death for cures.
> I follow straight without complaints or grief,
> Since if my scent be good, I care not if
> It be as short as yours.[87]

Conversely, of course, foul odour could indicate a lack of moral worth, as in Shakespeare's 'Sonnet 69':

> To thy fair flower add the rank smell of weeds
> But why thy odour matches not thy show
> The soil is this – that thou dost common grow.[88]

Ben Jonson, in turn, demolishes a critic of his poetry by stating emphatically that he stinks:

No man will tarry by thee, as he goes,
To ask thy name, if he have half his nose!
But fly thee, like the pest! Walk not the street
Out in the dog-days, lest the killer meet
Thy noddle, with his club; and dashing forth
Thy dirty brains, men smell thy want of worth.

On a more spiritual plane, fragrance, in the 'odour of sanctity' tradition, could stand for divine grace. George Herbert's religious poetry makes use of imagery of this sort. In 'The Odour: 2 Cor. 2.15' he describes the language of prayer in terms of fragrance.

How sweetly doth My Master sound! My Master!
As Amber-grease leaves a rich scent
Unto the taster:
So do these words a sweet content,
An oriental fragrancie, My Master.[89]

In 'The Banquet', the crucified Christ is compared to an aromatic wood which becomes more redolent when crushed:

But as Pomanders and wood
Still are good,
Yet being bruised are better scented;
God, to show how far his love
Could improve,
Here, as broken, is presented.[90]

Odour here signifies not just a byproduct, but the central value, the *essence* which gives meaning to its source and production, and to all life.

These poems, perhaps better than any of the other historical data, show the hold smell had on the contemporary imagination during this period. Rich and meaningful in popular culture and backed by classical and theological tradition, olfactory imagery served to evoke a whole range of emotions and ideas – from beauty to ugliness to moral worth to God.

THE OLFACTORY REVOLUTION

In the late eighteenth and early nineteenth centuries movements for sanitary reform began to grow in the cities of Europe. With the multiplication of factories and the rise in urban populations, the problem of waste and garbage disposal had become truly monumental. The need for reform of this sort was made more pressing by the cholera and typhus epidemics of the nineteenth century which were suspected of being spread by the odours of waste products.

Earnest reformers applied themselves to the task of recording in vivid detail the filth and stench of their cities in the hope that their writings would help bring about change. The British physician John Hogg, for example, decried the existence of slaughter-houses 'reeking with gore' in London and the attendant driving of huge herds of animals through the city:

> Whole trains of coaches, omnibuses, and wagons, are stopped by bullocks and sheep... often do the poor animals, overheated, and faint with thirst, rush towards a gutter of liquid filth, and drain it of its black and putrid contents, often do they drop and die in the streets from ill-usage and exhaustion, and frequently are they crushed and destroyed by the wheels of heavy-laden vehicles, and so the butcher's knife is cheated of its victim![91]

Apart from the herds of animals driven to be slaughtered, thousands of cows were kept in the city by dairies. Hogg writes that these establishments could be smelled from several streets away, and that 'it is not the "ambrosial breath" of the cow that is experienced ... but it is the filth that is accumulated in the sheds where the cows are so closely packed.'[92]

Another British physician, Hector Gavin, wrote a report on an olfactory tour he had made of a London suburb in which each street seemed more foul than the last, the only alleviation being an occasional flower garden. 'I could not remain to make notes of this place, so overpowering was the stench,' he writes of one stop on his tour, and of another, 'the stench was perfectly unendurable.'[93] Worst of all was the yard of a manure manufactory:

> To my right in this yard, was a large accumulation of dung, &c.; but, to the left, there was an extensive layer of a compost of blood, ashes, and nitric acid, which gave out the most horrid,

offensive, and disgusting concentration of putrescent odours it has ever been my lot to fall victim of.[94]

Such manure manufactories underline the fact that there was a profit to be made out of refuse. Dung, of course, had value as a fertilizer. It was also employed by tanneries to soften leather. Sugar refineries, in turn, made use of vast amounts of animal blood and parts in their processing. In *Les Misérables* Victor Hugo decries the 'loss of the hundred millions which France annually throws away', by letting potentially valuable manure be carried away to the sea.[95]

> Those heaps of garbage at the corners of the stone blocks, these tumbrils of mire jolting through the streets at night, these horrid scavengers' carts, these fetid streams of subterranean slime which the pavement hides from you, do you know what all this is? It is the flowering meadow ... it is perfumed hay, it is golden corn, it is bread on your table, it is warm blood in your veins, it is health, it is joy, it is life.[96]

This, of course, was the traditional farmer's perspective whereby the odours of excrement were tolerable, and even desirable, because they turned into the scents of harvest. In the ever-expanding cities of Europe, however, there were no nearby fields to fertilize, and so streets and rivers continued to float with sewage.

It was not only waste and its malodours which were considered dangerous by the sanitary reformers, however, but also the exhalations of living beings. Scientists, after observing animals writhe to death inside the vacuum of bell jars, had concluded that the circulation of air was essential to life. Without fresh air, therefore, the poor, crammed in their suffocating dwellings, would die of their own exhalations like animals in bell jars. Thus Gavin writes, for example:

> The air which is breathed within the dwellings of the poor is often most insufferably offensive to strangers. It is loaded with the most unhealthy emanations from the lungs and persons of the occupants – from the faecal remains which are commonly retained in the rooms – and from the accumulations of decomposing refuse which nearly universally abound ... In numerous instances, I found the air in the rooms of the

poor ... so saturated with putrescent exhalations, that to breathe it was to inhale a dangerous, perhaps fatal, poison.[97]

Once the problem was described, the difficulty lay in trying to rectify it. One major objection to keeping filth off the streets was that a great number of poor – street sweepers, scavengers, manure sellers, and so on – depended on it for their livelihood.[98] Indeed, in Paris in 1832, when attempts were made to improve the removal of rubbish, the poor rioted.[99] It was a question of environment versus employment, and when faced with this choice, many nineteenth century officials, businessmen and workers came down on the side of employment, as they often do over similar issues today. Clean streets were a luxury, they argued, jobs were a necessity.

With regard to human waste, networks of drains existed in the larger towns, but these were poorly made and inadequate. Again, there were reasons for delaying improvements. Individuals and communities were resistant to change and unwilling to spend money on expensive sewage systems. There was also the matter of all the door-to-door waste collectors who would be put out of work. In London, a legal issue was even made out of who had property rights over human waste – those who produced it, those who owned the property in which it was produced, or the state? Then again, waste being a very indelicate matter, there were many persons in that prim age who were unwilling to discuss it at all or even admit its existence.[100]

In the end, it took the increasing numbers of deaths from cholera epidemics to convince reluctant governments to institute measures of sanitary reform – house inspections, flush toilets, sewage systems, and so on. There was an olfactory impetus as well. Hot summers intensified urban stench until it became unbearable even for hardened city dwellers.[101] In London, for example, the summer of 1858 was so foul that it was suggested that Parliament be moved out of the city.[102] Instead, the great work of urban waste disposal began.

In the late nineteenth century the discovery was made that it was not smells that spread disease, but germs. However, since the germs which communicated diseases such as typhus and cholera could be found in waste products, the safe disposal of waste remained as important as before. As the network of drains and sewers spread, the olfactory ambience of European towns and cities slowly lost its excremental flair. With waste odours out of the way, the populace grew less tolerant of industrial stenches, and these too became subject to goverment control.[103] Foul odours were no longer considered an unpleasant but inevitable part of life; they were now an unacceptable affront to public sensibility, if not to public health, which could and should be eradicated.

This revolution in civic cleanliness was accompanied by a revolution in personal cleanliness. Baths, for instance, had reappeared in Europe in the eighteenth century. One important turnabout which made bathing acceptable and even desirable was that, whereas previously bathing had been thought to endanger one's health, it was now thought to be good for health. Body dirt, it was claimed, prevented perspiration and oil from being released by the skin, thus causing illness. As one late eighteenth-century French report put it: 'Major diseases ... most often occur when evacuations of the skin do not take place, nothing presenting a greater obstacle to this than body dirt and filth.'[104]

Hence, while before it had been thought necessary to leave one's skin unwashed so as to prevent it from being invaded by external effluvia, it now was argued that it was necessary to wash one's skin so as to allow it to release corrupt internal fluids. Scientific support was lent to these new beliefs by grotesque experiments in which it was observed that animals would slowly asphyxiate when their skin, coated with tar, was unable to breathe.[105] In the sixteenth century Francis Bacon had prescribed a bath routine which would carefully limit the 'breathing of the pores'; the emphasis now was on bathing in order to let one's skin breathe.

As the upper and middle classes, at first reluctantly, began to purify their bodies, their homes and their streets of dirt, they grew more conscious of the malodours of the working classes which did not. Among many in this latter group, the old standards and methods of personal cleanliness held good until the end of the nineteenth century. An English study on hygiene conducted in 1842, for instance, reports a labourer replying, when asked how often he washed, that: 'I never wash my body; I let my shirt rub the dirt off, my shirt will show that; I wash my neck and ears and face, of course.'[106]

Furthermore, the poor did not (and could not) separate the functions and odours of their households into discrete compartments – bedroom, bathroom, kitchen, dining room – as the moneyed classes did. Odours thus mingled indiscriminately in the crowded homes of the poor, increasing the revulsion felt towards them by the sensitized bourgeoisie, who had come to associate olfactory promiscuity with moral promiscuity. A Victorian perfumer writes, for example:

> Among the lower orders, bad smells are little heeded; in fact, 'noses have they, but they smell not;' and the result is, a continuance to live in an atmosphere laden with poisonous odours, whereas anyone with the least power of smelling retained shuns such odours, as they would anything else that is vile or pernicious.[107]

The olfactory reform of the poor was thus intimately linked with their moral reform. The new doctrine of cleanliness did eventually penetrate the working classes, due to the teaching of hygienic practices in the expanding school system, the amelioration of the living conditions of workers and the construction of public baths.[108] Even then, bathing was not necessarily, as we tend to think today, a purifying experience. George Orwell, for instance, recalled from the baths of his schooldays in the 1910s, 'the slimy water of the plunge bath', 'the always-damp towels with their cheesy smell', and 'the sweaty smell of the changing room'. He noted in conclusion: 'It is not easy for me to think of my schooldays without seeming to breathe in a whiff of something cold and evil-smelling.'[109] While the foul scents of Orwell's schoolboy baths are reminiscent of the bad-odoured old days, however, his hypersensitivity to them is characteristic of the new, sanitized, olfactory order.

Interestingly, this rise in personal cleanliness was accompanied by a decline in the use of perfumes. The most apparent reason for this would seem to be that, once bathing was an established practice, perfumes were no longer needed to mask unpleasant body odours. Nevertheless, there were a number of other factors influencing this shift as well. At the same time as washing with water was increasingly being judged healthy, perfumes were being stigmatized as unhealthy. No longer attributed any protective qualities by the medical profession, perfumes were instead deemed to clog the pores, or to enfeeble through their heavy vapours.[110] Indeed, for some, perfumes were almost as unhealthy as stenches. The great nineteenth-century sanitary reformer

Edward Chadwick, for instance, was of the opinion that 'all smell is disease'.[111]

Perfumes were therefore taken out of the pharmacy and relegated to the cosmetics counter, and their role was changing there as well. In the late eighteenth century, styles in clothes and cosmetics became more subdued, and the use of perfumes was likewise toned down. The French Revolution, with its revolt against aristocratic excess, furthered this trend towards sobriety. While the imagined corruption of the poor was associated with filth and stench, that of the aristocracy had its olfactory sign in heavy perfumes. The rising middle classes, in contrast, would find their niche in the safe middle ground of olfactory neutrality.

One important factor which linked perfume in particular to extravagance, was its ephemeral nature. Money that was spent on perfume literally evaporated, a process that represented the antithesis of bourgeois values of converting money into solid assets. Buying perfume was like scattering your money to the wind. Perfumes, consequently, no longer considered essential, entered the category of wasteful frivolity.

Scents served not only to mark differences of class during this period, but also of gender. Up until the end of the eighteenth century, perfumes had been extensively used by men and women alike. At that time, as has been noted, the use of perfumes declined. Whereas many men left off wearing scents altogether, however, women merely changed to lighter, floral fragrances.

Furthermore, traditionally the same perfumes had been used by both men and women. It is related of George IV of England (who reigned from 1820 to 1830), for example, that he first encountered what was to be his favourite scent for his own person worn by a princess at a ball.[112] By the late nineteenth century, however, certain scents – in particular, sweet floral blends – were deemed exclusively feminine, while other, sharper, scents were characterized as masculine. The burgeoning perfume industry capitalized on these trends by creating and promoting perfumes specifically for women, and, to a much lesser extent, others, marketed as aftershaves or colognes, for men.

What were the reasons for this olfactory divide of the sexes? The typing of perfumes as frivolous, for one, made them suitable only for 'frivolous creatures', and in nineteenth-century society that meant women. Sweet, floral fragrances were considered feminine by nature because, according to the gender standards of the day, 'sweetness' and 'floweriness' were quintessentially feminine characteristics. If the flower garden was classified as a female domain, however, the woods were typed as a male one, making 'woodsy' scents, such as pine and cedar, an acceptable alternative for men. Properly, none the less, men were expected to disdain all such olfactory artifice and smell only of clean male skin and tobacco. This emphasis on the olfactory difference between men and women was part of a general cultural insistence at the time that the sexes appear in all ways to be different.[113]

It was not just perfume which became feminized in the nineteenth century, however, but the whole sense of smell. Beginning with the Enlightenment, smell had been increasingly devalued as a means of conveying or acquiring essential truths. The odour of sanctity was no longer an influential concept, nor were smells thought to have therapeutic powers. Sight, instead, had become the pre-eminent means and metaphor for discovery and knowledge, the sense *par excellence* of science. Sight, therefore, increasingly became associated with men, who – as explorers, scientists, politicians or industrialists – were perceived as discovering and dominating the world through their keen gaze. Smell, in turn, was now considered the sense of intuition and sentiment, of home-making and seduction, all of which were associated with women. It was maps, microscopes and money on the one hand, and pot-pourris, pabulum and perfume on the other. Significantly, however, smell was also the sense of 'savages' and animals, two categories of beings who, like women, were deprecated and exploited by contemporary Western culture.[114]

The upheaval of the First World War further altered the perception of smell by causing many of the qualities which had come to be associated with it – sentimentality, intuition, nostalgia – to be considered obsolete and even ridiculous in the fast-paced and hard-nosed modern world. Pot-pourris had no place in the functional twentieth-century home. Flower shows could not compete with the cinema. It was at this time that the modern olfactory era began in the West, an era characterized by the widespread deodorization of public and private space; the restriction of perfumes to personal use, often on special occasions only and primarily by women; and a general devaluation of, and inattention to, olfactory power and meaning.[115]

SCENTS OF IMAGINATION

As odours were chased out of mainstream culture in the nineteenth century, they were taken up and transformed into literary signs by an avant-garde group of writers and poets.[116] The reasons for this literary flowering of scent were various. Certain writers, such as Victor Hugo, Honoré de Balzac and Emile Zola, set out to depict the olfactory landscapes of their novels as graphically as the sanitary reformers detailed those of the streets and cities they wished to cleanse. In *Père Goriot*, Balzac, for example, describes the odour of a boarding-house in meticulous detail:

It smells stuffy, mouldy, rancid; it is chilly, clammy to breathe, permeates one's clothing; it leaves the stale taste of a room where people have been eating; it stinks of backstairs, scullery, workhouse. It could only be described if some process were invented for measuring the quantity of disgusting elemenatary particles contributed by each resident, young or old, from his own catarrhal and *sui generis* exhalations.[117]

A revolting account indeed, yet realism demanded no less.

Apart from its ability to spice up a narrative, smell carried with it a long history of cultural values. Odour might have been divested of any real power in the nineteenth century, but its symbolism remained intact, making smell an eminently useful literary device for creating a moral atmosphere at once forceful and indirect. Thus in *Nana*, through a litany of suggestive scents, Zola is able to convey the seductive yet oppressive atmosphere of the backstage dressing room.

Muffat was beginning to perspire: he had taken his hat off. What inconvenienced him most was the stuffy, dense overheated air of the place, with its strong haunting smell ... In the passage the air was still more suffocating, and one seemed to breathe a poisoned atmosphere ... High above him there was ... a banging of doors, which in their continual opening and shutting allowed an odour of womanhood to escape – a musky scent of oils and essences mingling with the natural pungency exhaled from human tresses. He did not stop. Nay, he hastened his walk: he almost ran, his skin tingling with the breath of that fiery approach to a world he knew nothing of.[118]

It was as though, once odours were disempowered by science, they were free to be empowered by the imagination.

The Symbolist writers – Baudelaire, Mallarmé, Verlaine, Wilde, Machado, Rilke, among others – found that scent, with its nuances of sin and sanctity, exoticism and seduction, provided a rich imagery for their exploration of themes of sensual decadence and transcendence. The dream-like atmosphere the Symbolists sought was evoked by references to odours, themselves dream-like in their ephemerality and formlessness. The intrinsic formlessness of smell, in turn, made it an apt literary metaphor for the formlessness of emotions. Sadness, joy, desire, horror, hope, are revealed as persistent yet nebulous scents in Symbolist writing:

> And in your garments that exhale your perfume
> I would bury my aching head
> And breathe, like a withered flower,
> The sweet, stale reek of my love that is dead.[119]

Literary odours thus served both Realists, who employed them to give their writings the pungent scent of truth, as well as to make moral statements, and Symbolists, who transformed them into lush, emotion-laden images to convey an essence of dreams (or nightmares).

The evocative nature of smell led certain nineteenth-century writers to meditate on the existence of a primal language of smell. Oscar Wilde, for example, has his protagonist explore the associations between odours and emotions in *The Picture of Dorian Gray*:

> He saw that there was no mood of the mind that had not its counterpart in the sensuous life, and set himself to discover their true relations, wondering what there was in frankincense that made one mystical, and in ambergris that stirred one's passions, and in violets that woke the memory of dead romances, and in musk that troubled the brain.[120]

In Huysmans' classic novel of aesthetic excess, *Against Nature*, the protagonist surrounds himself with perfumes in an attempt to master the 'syntax of smells':[121]

> Little by little the arcana of this art, the most neglected of them all, had been revealed to Des Esseintes, who could now decipher its complex language that was as subtle as any human

tongue, yet wonderfully concise under its apparent vagueness and ambiguity.[122]

This olfactory language was often expressed as being interrelated with other sensory codes. From the classical period through to the Renaissance it had been widely thought that there was a fundamental system of universal correspondence by which a certain planet was linked with a certain colour, perfume, gemstone, plant, and so on. Picking up on this tradition (by way of the writings of the Swedish mystic Swedenborg), the Symbolists explored the notion of a sensory correspondence by which sounds would be associated with scents or tastes with colours. 'All scents and sounds and colors meet as one,'[123] writes Baudelaire in his poem 'Correspondences':

> Perfumes there are sweet as the oboe's sound,
> Green as the prairies, fresh as a child's caress.[124]

Huysmans' protagonist, in turn, transposes poems by Baudelaire into perfume, creating 'aromatic stanzas'.[125] During the same period perfumers were suggesting that scents could be correlated with the musical scale, with individual fragrances constituting notes, and blends harmonies (a terminology still in use today in perfumery).

Making odour the subject of literary discourse was not simply a matter of aesthetics, however, but also often a political statement. The suppression of certain odours – excrement, heavy perfumes, and so on – by mainstream society, for example, made giving literary voice to them a means of rebelling against the constraints of that society. By referring to 'unspeakable' scents, Verlaine, Rimbaud and others flaunted their unwillingness to be bound by the sterile social conventions of the bourgeoisie.

The literary championing of smell also served in certain cases as a protest against the increasing commercialization of modern life. In England, for example, writers such as William Morris decried the emphasis on show over scent in modern flowers and gardens, seeing in the process a replacement of traditional spiritual values with capitalist values of mass production and conspicuous wealth.[126]

Finally, perhaps because the growing deodorization of society was creating a nostalgia for lost scents, smell became intimately associated with memory in the nineteenth century. Consequently, writers who wished to evoke a nostalgic atmosphere of bygone days often used olfactory references to set the mood. The most notable examples of this are found in Proust's *Remembrance of Things Past*, written in the early twentieth century. In that monumental novel the narrator's reminiscences are triggered by the taste and smell of a madeleine dipped in tea, a treat which his aunt used to give him as a child. These reminiscences, in turn, are scented with the odours of childhood:

> smells changing with the season, but plenishing and homely, offsetting the sharpness of hoarfrost with the sweetness of warm bread, smells lazy and punctual as a village clock, roving and settled, heedless and provident, linen smells, morning smells, pious smells.[127]

When all else is gone, the narrator states, smell and taste, 'bear unflinchingly, in the tiny and almost impalpable drop of their essence, the vast structure of recollection'.[128] As the twentieth century progressed, however, that precious drop of essence would increasingly evaporate unattended.

SMELL AND SCIENCE

The interest in smell shown by writers of the nineteenth century was also manifested by scientists. Whereas literature tended to glorify smell, however, science tended to depreciate it. Already in the sixteenth century, René Descartes had made it clear that the sense of science was to be sight and this position was strengthened in the following centuries. Smells, so hard to measure, name or recreate, were undoubtedly among the least accessible sensory stimuli to the methods of science.

None the less, during the eighteenth and nineteenth centuries, when science was avidly exploring the former domains of religion, folklore and alchemy, odours were for a time an important subject of scientific investigation and discourse. Human odours, for example, were enthusiastically, if not very reliably, classified by sex, race, age, diet and even hair colour (brunettes were said to smell pungent and blondes musky) by the scientists of the period.[129]

It was the odours of putridity, however, which captured most of the scientific interest directed towards smell, as Alain Corbin has amply documented in his book on the perception of smell in eighteenth- and nineteenth-century France. This was due to the general belief in stench as a major source of disease. Certain scholars thus devoted themselves to studying the odours of street filth; others investigated the scents exhaled by prison or hospital walls; still others, the odours produced by the decomposition of corpses or excrement. Dedicated physicians and chemists, surveying the stenches of polluted rivers, produced descriptions of fetidity which rival the olfactory poetics of Ben Jonson's 'The Famous Voyage', but within the context of a scientific enquiry.[130]

All this came to an end in the late eighteen hundreds. Aromatics had already been dismissed by science as serving only to mask, not transform, foul odours. Now, Pasteur's discovery that most familiar diseases are caused by germs led scientists to conclude that foul odours themselves were not agents of illness, but merely rather unimportant byproducts. The medical community left smells behind and moved on to microbes. In the scientific paradigm of the universe, odours had become inessential.[131]

Paralleling and informing the scientific discourse on smell, was the philosophical discourse. According to the dominant philosophic trends of the Enlightenment, smell offered neither a significant means of acquiring knowledge nor of aesthetic enjoyment. Condillac, in his *Treatise on the Sensations*, for example, remarked that 'of all the senses [smell] is the one that seems to contribute the least to the operations of the human mind'.[132] His contemporary Kant agreed, relegating smell to the dustheap of the senses:

> To which organic sense do we owe the least and which seems to be the most dispensible? The sense of smell. It does not pay us to cultivate it or to refine it in order to gain enjoyment; this sense can pick up more objects of aversion than of pleasure (especially in crowded places) and, besides, the pleasure coming from the sense of smell cannot be other than fleeting and transitory.[133]

Such an all-out condemnation of smell reeks of a major sensory repression. Yet, as the scientists and psychologists of the nineteenth and early twentieth centuries would argue, the suppression of the sense of smell was one of the defining characteristics of 'civilized man'. Darwin had postulated that humans lost their acuity of smell in the process of evolving from animals.[134] The marginalization of smell in human society, therefore, appeared necessary for evolutionary and cultural progress, while any attempt to cultivate smell would signify a regression to an earlier, more primitive state. Freud and previously Herder held that smell had given way to sight when the human species began to walk upright, removing the nose from the proximity of scent trails and increasing the visual field. Since, according to Freud, individuals repeat the process of evolution in their psychological development, as a person matures, the revelling in odour of the infant should likewise give way to visual pleasures. Adults who continue to emphasize the olfactory are hence arrested in their psychological development.[135]

At the turn of the twentieth century, Havelock Ellis wrote extensively on the pyschology of smell. He concluded, as had others in his field, that:

> The perfume exhaled by many holy men and women ... was doubtless due ... to abnormal nervous conditions, for it is well known that such conditions affect the odor, and in insanity, for instance, the presence is noted of bodily odors which have sometimes even been considered of diagnostic importance.[136]

The 'odour of sanctity' occurring after death, in turn, was attributed by Ellis to a confusion with the *odor mortis*. As for the reputed olfactory sensitivity of many saints, he noted that 'smell and taste hallucinations appear to be specially frequent in forms of religious insanity'.[137] Not only could insanity be productive of abnormal odours and olfactory delusions, odours could also be productive of insanity. Ellis notes that 'dealers in musk are said to be specially liable to precocious dementia.'[138]

The olfactory imagery in the works of many nineteenth-century writers was also explained by Ellis in terms of a pyschological disorder:

> It is certain also that a great many neurasthenic people ... are peculiarly susceptible to olfactory influences. A number of eminent poets and novelists – especially, it would appear, in France – seem to be in this case.[139]

The German writer Max Nordau stated this position more strongly in his book, *Degeneration*. In this work he condemns Zola, for example, for presenting characters in his novels not as 'normal individuals, viz., in the first instance as optical and acoustic phenomena, but as olfactory perceptions'.[140] Sight and hearing are thus established as the acceptable media for the perception of others, while smell becomes abnormal. Nordau rhetorically asks: 'Why should the sense of smell be neglected in poetry? Has it not the same rights as all the other senses?'[141] He responds by saying that individuals cannot set themselves against 'the march of organic evolution':

> The underdeveloped or insufficiently developed senses help the brain little or not at all, to know and understand the world ... Smellers among degenerates represent an atavism going back, not only to the primeval period of man, but infinitely more remote still, to an epoch anterior to man.[142]

Not all contemporary scientists agreed with this position, however. In fact, there were some who went to the other extreme and elaborated theories of smell almost as mystical as the olfactory revelations of the saints. One such, August Galopin, in a book entitled *Le parfum de la femme*, asserted that

> The purest marriage that can be contracted between a man and a woman, is that engendered by olfaction and sanctioned by a common assimilation in the brain of the animated molecules due to the secretion and evaporation of two bodies in contact and sympathy.[143]

Such aromaphiles were blowing against the wind, however, and it is not their work, but that of Freud and Ellis which survived to influence posterity.

The late nineteenth- and early twentieth-century scholars of olfaction were aware of the continuing presence of traditional olfactory beliefs in many rural European communities. These they tended to dismiss, however, as curious but archaic customs. Similarly, there was significant interest among anthropologists in describing the olfactory practices of non-Western cultures; but this was not with the purpose of elevating the sense of smell, but rather of devaluing the peoples who so elaborated it. A higher olfactory consciousness in non-European cultures was taken as one more proof of their lower status on the evolutionary scale of civilization.[144]

Such comparative studies of olfactory and other sensory practices lost favour after the Second World War. The odours and flavours of other cultures, it was thought, were mere 'picturesque' details that belonged more in travelogues than in serious anthropological literature. Furthermore, studying the role of smell among Third World peoples seemed to smack too much of the unsavoury racist theories of the nineteenth century which associated smell with savagery. Just as the anthropologists of that time had sought to denigrate non-Europeans by bringing out their reliance on the 'lower' sense of smell, modern anthropologists sought to render them as 'civilized' as Europeans, by deodorizing their cultures.

By the mid-twentieth century, anthropologists, with a few exceptions,[145] would stop even noticing cultural differences of smell. There was now no apparent alternative to the olfactory illiteracy of the modern West. Whatever (marginal) role smell played in the West was (and is) assumed to be the same the world over. However, as we shall learn in the following chapters, recent cross-cultural research on smell increasingly shows how far this is from being the case. The olfactory condition of the modern West is not a definitive model for the role of smell among all peoples at all times, but simply the result of certain very particular historical and cultural circumstances, a result as subject to continuing shifts and transmutations as odour itself.

Part II

Explorations in olfactory difference

Chapter 3

Universes of odour

What is it like to live in a society where time is conceived of as a succession of odours? In what ways have smells been used to classify people, animals and plants in different cultures? What happens when the olfactory codes of a society are broken and smells which should be kept separate mix and mingle? In this chapter we shall be entering into a variety of cultural universes of odour, from Africa to the Amazon and from China to New Guinea. What interests us are the 'osmologies', or classificatory systems based on smell, which are used to order the world by the peoples of those lands.[1] Our account begins in the Andaman Islands, which lie off the coast of Burma in the Bay of Bengal.

SCENTED CALENDARS: THE AROMAS OF TIME

In the jungles of the Andaman Islands, as one after another of the trees and climbing plants come into flower, it is possible to recognize a distinct succession of odours. The Andaman Islanders have constructed their calendar on the basis of this cycle, naming the different periods of their year after the fragrant flowers that are in bloom at different times. Their year is thus a cycle of odours; their calendar, a calendar of scents.[2]

To the Andaman Islanders, aromas are vital energies, and smell is power. For example, the efficacy of a magic talisman or medicinal plant is identified with its odour. Similarly, each floral season is believed to possess its own particular kind of 'aroma-force'. The succession of these aroma-forces is thought to yield the different plants and fruits that appear successively with the changing seasons. These seasonal aromas, therefore, constitute the fundamental generative agencies of nature.[3]

By eating the different fruits of the forest, humans internalize the vital energy of the scents thought to produce them. Honey, a favoured forest product, is particularly associated with the scents of the seasons, as it receives its distinctive aroma and flavour from the flowers in bloom at the time of its production. As the seasons of the year change, therefore, the Andaman Islanders enjoy a succession of aromatic honeys – making the passage of time sweet indeed.

Aroma-forces are thought to exert an influence over the human life cycle just as they order the life of the jungle. For example, a girl is given the name of the fragrant flower in bloom at the time of her first menstruation, a time referred to as her 'blossoming time'. She keeps this flower-name until she bears her first child, or in local parlance her 'first fruit', after which she goes back to being called by her birth names.[4] The successive aroma-forces of the jungle thus regulate the human reproductive cycle in the same way as they order the growth of plants.

Among the Dassanetch, a farming and cattle-herding people of Ethiopia, time is also ordered by a succession of odours. Here it is not floral scents that alternate, however, but rather the smells of burning and decay characteristic of the dry season, on the one hand, with the fresh smells of new plant growth that arise during the rainy season, on the other. The Dassanetch year consists of two dry seasons and two wet seasons. In the dry seasons, fields are burnt to clear them of old growth, filling the air with acrid smoke. The dry seasons are also a period of decay, as plants wither and die and overly ripe fruits rot. All of these unpleasant smells are said to rise up to the sky where they are absorbed by the clouds and dissipated. In the wet seasons, rain comes down from the clouds, enabling new grass to grow in the fields and the plants to revive and bloom, making the world smell sweet and fresh again.[5]

The calendar of the Dassanetch is, essentially, one in which the stagnant stench of the dry season is followed by the refreshing fragrance of the wet season. In other words, the Dassanetch time cycle is composed of smells of destruction followed by smells of creation. However unpleasant the Dassanetch might find the odours of burning and decay, they recognize that both bad and good smells are necessary to the rhythm of time and life, and that one olfactory season prepares the way for the next. Indeed, the only scent truly abhorred by the Dassanetch is that of fish, believed to live outside the cycles of nature in the seemingly seasonless world of water. As a literally timeless and unnatural scent, the odour of fish is an abomination, having no place in the scent calendar of the Dassanetch.

To breathe in the odours of nature for the Andaman Islanders, or the Dassanetch, is to take into oneself the *essence* of time and life. It is to feel the vital energy of the cosmos acting through and around one, in an endless cycle of growth, decay and renewal.

SCENTED MAPS: SMELLSCAPES

Just as odours exist in time and change with time, so do they exist and change within space. In the Andaman Islands, for example, when the species of *Sterculia* flower called *jeru* comes into blossom, 'it is almost impossible to get away from the smell of it except on the sea shore when the wind is from the sea'.[6] Such potent floral scents make the Andamans 'islands of fragrance', surrounded by the salty smell of the sea. This opposition between the air of the jungle and the sea provides the Andamanese with an olfactory definition of the space in which they live. That space is further differentiated by the more localized odours of villages, dens of animals, different plant zones, and so on. Together these configurations of odours constitute the olfactory landscape, or 'smellscape', of the Andaman Islands.[7]

Such a 'smellscape' is obviously not a fixed structure, but rather a highly fluid pattern that can shift and change according to atmospheric conditions. Perhaps because of the importance their culture attaches to smell as a means of ordering the world, the Andamanese conceive of space itself not in the way most people do in the West – as a static area within which things happen, but more as a dynamic environmental flow. Consider, for example, the space of the village. This space is experienced and conceptualized as fluctuating over time: it can be more expansive or less, depending on the presence in the village of strong-smelling substances (such as pig's meat), the heat, the strength of the wind, and so on.[8]

The Andaman Islanders are deeply concerned with controlling the olfactory perimeters of the village, for smell is thought to be used by the spirits who traverse the islands to determine the whereabouts of their human counterparts. Contact with spirits outside ritual contexts is believed to be highly dangerous, and therefore to be avoided. When the Islanders wish to conceal their presence from the spirits, as they do most of the time, they accordingly make every effort to reduce and mask their own olfactory space; when they wish to attract the spirits, they instead attempt to increase it.[9]

It would seem that the importance of smell as a means of spatial orientation and localization is often heightened in tropical forest environments. This is perhaps because odours abound in the dense forest atmosphere while vision is restricted. Consider the following description of a New Guinea rainforest by the anthropologist Gilbert Lewis:

> Although it is usually easy to walk through the forest, there are no perspectives, no open views: a companion becomes lost to sight among leaves, stems, shadows and trunks when he has walked twenty yards away. The light is dimmed and greenish ... The air is still and it smells musty ... Occasionally one passes through a path of unmoving air faintly scented by some plant like honey-suckle; one passes transient smells, of humus, of moist rotting wood or bruised fruits.[10]

The native inhabitants of New Guinea are, in fact, extremely alert to the olfactory cues of the rainforest. The Umeda people of the Sepik River region, for instance, are said to be 'brilliant' at detecting the faintest trace of smoke from a campfire in the depths of the forest, or at spotting where a cuscus (a pungent-smelling possum) might be concealed in the forest canopy.[11] Interestingly, whereas in the West sight is considered *the* distance sense, smell often outdistances sight in the experience of forest dwellers like the Umeda. They know that smell can give them knowledge of things hidden to the eye.

Another example of smells being used to order the experience and understanding of space is provided by the Desana of the Amazonian rainforest of Colombia. The Desana say that the territory inhabited by a tribe is permeated by *máhsa serirí*, a term that means both 'tribal odour' and 'tribal feeling' or 'sympathy'. In the same way that certain animals mark out their territories by scent, the Desana hold that tribal territory is marked out by the scent trails laid down by the people who live there. Each tribe is deemed to emit a unique odour, with the result that each tribal territory has a characteristic scent. This scent is supposed to remain discernible even where there is no immediate sign of human habitation. Thus, when travelling to other regions, the Desana continually sniff the air and comment upon the distinct odours of the different tribes that inhabit those regions. In fact, the Desana call themselves *wira*, which means 'people who smell' and refers to both the emphasis they place on olfaction as a way of knowing, and their particular tribal body odour.[12]

The different areas of the jungle – from deep to open forest – are also characterized by distinctive odours, according to the Desana, and the odour of the animals living in each area is thought to be conditioned by the smell of their surroundings. Thus, deep forest animals, such as the peccary and jaguar, are said to give off the unpleasant musky smell of the forest depths, while animals that live in more open spaces such as clearings, including various kinds of rodents (agouti, paca), are thought to have the pleasant, sweet smell of the open forest.[13]

As the animals move through their environments they lay down scent trails, called 'wind threads' by the Desana, both marking out their own territories and enabling other animals and humans to track them down. 'Wind threads' are also emitted by the different plants and fruits of a region, which, in turn, can be followed by humans and animals to their source. For the Desana, therefore, the smellscape of their environment consists of a variety of distinct olfactory zones criss-crossed by the different scent trails of the people, animals and plants which live in them.

ODOUR TYPES: OLFACTORY CLASSIFICATION

It is apparent from the preceding discussion that the organization of olfactory time and space is closely bound up with the typing of people, animals and plants by odour. We have seen this in the case of the Desana, whose system of olfactory classification – or 'osmology' – is very rich indeed and merits further study. As mentioned, the Desana believe that each tribal group has its own characteristic odour. This odour is said to be due in part to heredity and in part to the kinds of foods consumed. Thus, the Desana, who are hunters, are said to exude the musky smell of the game which they eat. Their neighbours, the Tapuya, on the other hand, live by fishing and are thought to smell of fish. The nearby Tukano are agriculturalists and they, in turn, are said to smell of the roots, tubers and vegetables which they grow in their fields.[14]

The Desana hold that men and women also have different characteristic odours. In general, men are said to smell of meat and women of fish. These basic smells can vary, however, among tribal groups. Desana women, for example, are said to smell of a very pungent kind of ant, or of a certain large worm that lives in the forest trees. For this reason, Desana women are sometimes called 'worm-odour women'. During rituals, such ants or worms can symbolize Desana women by their smell.[15]

While Desana olfactory categories serve to distinguish different species of animals and plants, they can also cut right across species divisions to create groups with nothing more in common than a similar smell. One such grouping, for example, could contain a variety of fruits, trees, birds and mammals, as well as humans. Such groupings are not simply arbitrary collections of fundamentally unlike entities, however, for every odour category possesses its own cultural value and potency, according to Desana notions. The odours of both black brocket deer and palm trees, for example, are associated with – and may therefore convey – male power and fertility. Virtually all of the fundamental ideas and values of Desana culture are 'odorized' in this way.[16]

It should be noted that the Desana entertain a somewhat larger concept of odour than one finds in mainstream Western culture, for they believe that smells are apprehended by the whole body, not simply the nose.[17] Smells, moreover, are the medium through which the values of the Desana moral code are rendered sensible, as in the above example of the odour of the black brocket deer standing for male power and fertility. This larger conception of odour explains why the Desana system of olfactory classification is not limited to dividing up elements of the environment into recognizable physical categories, the way a zoological or botanical classification would. The Desana taxonomy instead aspires to group together entities on the basis of their moral significance, so as to continually remind people of the basic ideals and energies which govern the Desana cosmos.

Another highly intricate South American osmology is the system of olfactory classification employed by the Suya Indians of the Mato Grosso region of Brazil. The Suya have three principal odour classes: bland-smelling, pungent-smelling, and strong-smelling (see Table 1). In the bland-smelling class the Suya place adult men, certain small mammals and birds, most fish, and innocuous plants. In the pungent-smelling class are placed old men and women, certain large mammals, the macaw, a few amphibians, and most medicinal plants. In the strong-smelling class are women, children, carnivorous mammals and birds, and harmful plants. As regards the cultural values attached to these classes, the stronger-smelling the class, the more potentially dangerous its members are deemed to be to human society.[18]

Adult men, the dominant group in Suya society, are considered bland-smelling because the male community of the men's house constitutes the ideal society in Suya culture. Women are assigned a strong odour because they are said to distract men from their ideal male community and because their fertility associates them with the asocial forces of nature. Old women are in the pungent, rather than strong-smelling class because they are deemed no longer attractive to men and no longer fertile. Old men, in turn, are said to be pungent-smelling because the relaxation of cultural restrictions in old age makes old men more 'natural' and less social. Similarly, children are classified as strong-smelling because they are as yet unsocialized and thus disruptive to the social order. [19]

Table 1: Olfactory classification system of the Suya Indians of Brazil

1. Bland-smelling	adult men small mammals and birds, most fish innocuous plants
2. Pungent-smelling	old men, old women large mammals, macaw, certain amphibians medicinal plants
3. Strong-smelling	adult women, children carnivorous mammals and birds harmful plants

This classificatory system, like that of the Desana, is heavily invested with cultural values. The supposed bland odour of men, for example, places them at the top of the social scale, while the alleged strong odour of women places them at the bottom. Odour, thus, serves not only to classify individuals but also to rank them. It is by reference to odour that people are situated within the Suya social order.

The Bororo, who inhabit the same region of Brazil as the Suya, have an equally extensive, although differently configured, system of olfactory typing. The Bororo distinguish eight principal odour classes, to one or another of which virtually everything in the world can be assigned. At one end of the classificatory system is *jerimaga*, a musky, rotten smell. Included in this category are skunks, vultures, mulberries and quince. At the other end is *rukore*, a sweet smell. Ducks, flamingoes, fragrant flowers and corn are examples of this category.[20]

The putrid *jerimaga* odour is said to be the characteristic smell of *raka*, which is the life force. This life force is associated with organic fluids, especially blood, milk and semen. Each person is born with a fixed amount of *raka* which is expended throughout life by physical activity. When one's supply of *raka* runs out, one dies. Men are said to be more prone to waste *raka* than women (except when the latter are menstruating or giving birth), and are therefore more likely to smell of *jerimaga*. While obviously essential to life, *jerimaga* is also highly dangerous, for it is considered to be a powerful agent of transformation. The Bororo therefore are careful to guard themselves against exposure to *jerimaga* and, when exposed, take particular care to wash themselves so as to lessen its effects.[21]

The sweet *rukore* smell, on the other hand, is thought to be the characteristic odour of the soul and is associated with the breath and the winds. After death, when one has lost all of one's *jerimaga*, one becomes pure *rukore*, pure soul. One then leads a spiritual existence, returning now and then to one's village as a cold gust of wind to be offered *rukore* food, such as boiled vegetables and sweet water, by one's survivors.[22]

Among the Bororo, the two basic smells, putrid and sweet, signify the two basic cosmic forces: life and spirit. This simple olfactory division provides the foundation on which the whole elaborate edifice of Bororo beliefs and practices concerning the body, the social and natural orders and the spirits, is constructed.

Let us now consider an African osmology – specifically, the system of olfactory classification of the Serer Ndut of Senegal. The Serer Ndut have five basic odour categories: urinous, rotten, milky or fishy, acidic or acrid, and fragrant (see Table 2). Of these five categories, only the last is thought to be agreeable. Included in the urinous category are plants used as diuretics, crushed squash leaves, monkeys, horses, dogs, cats, and ethnic groups that have a reputation for not washing, such as Europeans. Included in the rotten category are creepers growing on wet ground, mushrooms, pigs, ducks, camels and cadavers. The milky or fishy category includes goats, cows, antelopes, jackals, fish, frogs, certain neighbouring tribes and nursing women. Placed in the acidic class are certain trees and roots, tomatoes, donkeys and spiritual beings. In the fragrant category are flowers, limes, peanuts, raw onions and clean and perfumed humans, such as the Ndut and their neighbours, the Bambara.[23]

Table 2: Olfactory classification system of the Serer Ndut of Senegal

1. Urinous	Europeans monkeys, horses, dogs, cats plants used as diuretics, squash leaves
2. Rotten	cadavers pigs, ducks, camels creeping plants
3. Milky or fishy	nursing women, neighbouring tribes goats, cows, antelopes, jackals, fish, frogs
4. Acidic	spiritual beings donkeys tomatoes, certain trees and roots
5. Fragrant	Serer Ndut, Bambara flowers, limes, peanuts, raw onions

At first glance, it may be difficult to see any logic to the Serer Ndut osmology. How can spiritual beings and tomatoes belong to the same class? Why would milky and fishy odours be grouped together? Why are Europeans considered urinous? The seeming eccentricity of the Serer Ndut osmology is not unique, however, for all olfactory classifications are equally arbitrary. Odours have consistently defied attempts at rational (or 'objective') classification, and probably always will.[24]

It is precisely because they are so value-laden, however, that osmologies are so revealing of the essential preoccupations of a society. Olfactory classification systems do possess a sense, a logic, but that logic is local rather than universal. They can only be studied in context, and to understand them one must adopt the perspective of the other. The Serer Ndut classification, therefore, *is* logical, but its logic can only be grasped within the context of Ndut culture.

Europeans, for example, are accustomed to thinking of 'the natives' as unwashed, and themselves as clean and innocuous-smelling. From the standpoint of the Serer Ndut, however, the reverse is true: Europeans are foul and the Serer Ndut clean. Thus, a Serer Ndut mother will warn a child who dislikes being bathed: 'You're going to smell urinous like whites!'[25] This particular categorization, while possibly indicative of different hygienic practices, more fundamentally reflects the fact that the odour of the 'other' will usually make a stronger impression on us than the odour of our own kind. In the experience of the Serer Ndut, the categories of self and other are the reverse of what they are for Europeans, hence the reversal (from a Western perspective) in the assignment of good and bad olfactory identities.

The fact that in the Serer Ndut osmology the smell of onion is classified as fragrant is another point which may give the reader pause. Essence of onion could never enjoy success as a perfume in the West, although it does in Senegal and elsewhere in Africa. This example underlines the fact that the categories of the fragrant and the foul are not given in nature, but rather derive from culture. There are no natural likes or dislikes in matters olfactory.

The grouping of the odours of milk and fish in one category obeys an internal logic as well. What is essential to understand here is that the Serer Ndut's experience of milk and (dead) fish is that they both turn rancid very quickly in the heat. Therefore, it is the *rankness* of stale milk and fish that unites them in one odour category for the Serer Ndut.

Finally, there is the mystery of why spiritual beings should be considered acidic or acrid-smelling by the Serer Ndut. The explanation for this association comes from the fact that the Serer Ndut employ the acrid smoke of certain acidic-smelling plants to chase away snakes, which are associated with the spirit world. The acidic odour of some of these plants is, indeed, so strong that it can cause a person to faint. Such odours may be used by the Serer Ndut to assist a person who is in a coma to die by chasing the spirit out of the body. These associations are apparently so potent that they have had the effect of permanently scenting the Serer Ndut spirit world with an acidic and acrid odour.

Olfactory typologies, such as those we have been considering, divide up the natural and social universe along smell lines. They represent a culture's way of making sense of the world through scents, or its 'world-scent', if you will.[26] In the following section, we shall explore how such 'world-scents' operate in the particular context of one of the most basic domains of culture: food.

EDIBLE ODOURS: THE SMELL OF FOOD

From the curries of India to the chillies of Mexico, the scents and tastes of the traditional foods and sauces of different societies serve as an important means of cultural differentiation. The composition of such 'national dishes' reflects the food resources of the region. These dishes also, however, depend on cultural notions of what constitutes 'food'. Thus, for example, to a vegetarian Hindu, the roast beef dinner of an Argentine gaucho would seem morally and physically repugnant, while the gaucho, in turn, would dismiss a salad of greens as inedible 'grass', fit only for cattle.

In many cultures, the odours of potential foodstuffs are an important factor in whether they will be classified as edible or inedible. Among the Bororo, for example, the safest foods (suitable for babies and invalids) are deemed to be those with little odour or with a sweet odour, such as rice and corn. Foods, primarily meat, that are thought to have a putrid smell, are considered dangerous to the health. These foods are rendered edible by being boiled until none of the dangerous odour remains.[27] In all these cases, edibility does not depend solely on whether the food item is physically safe to eat (i.e. not poisonous or indigestible); it depends at least as much, if not more, on the *cultural values* associated with that food.

However culturally based (and therefore arbitrary or 'social') the reasons for avoiding certain food odours may be, the reactions of individuals to these stigmatized odours tend nevertheless to be highly personal and visceral. Consider, for example, the following statement made by an Uduk woman of Sudan about her response to the smell of pig meat. (The pig is an animal regarded with some ambivalence by the Uduk, due in part to the influence of Islam.)

> I can eat pig because people give me medicine, and then I can eat it. I used to refuse pig, because when it is roasted, *caah!* Its smell comes rushing into you, to the Liver there, and your Liver goes *guug!* And you vomit, *woog!*[28]

An extensive system of classification of foods by odour is found in the culture of the Colombian Desana. According to the Desana, the odour of a food determines how it should be processed. Game and certain fish, such as catfish, are said to have a foul, musky smell of menstrual blood. These animals must be smoked to alter their odour and then boiled, before they are eaten. Meat from other animals, and most vegetables, with less potent odours, are boiled directly. The most harmful food odour is that of burning fat, which is compared to the smell of rutting animals. Its undesirability for the Desana, who are perpetually concerned with controlling and channelling human sexuality into socially sanctioned forms, lies in its strong sexual overtones. In order to avoid producing this sexually charged odour, meat is rarely roasted and never fried by the Desana.[29]

The best food is that deemed to be in a properly 'ripe' state, with ripeness signifying not simply maturity, but also odour and other sensory attributes, as well as cultural acceptability. As a rule Desana prefer to eat rodents, such as paca, agouti and armadillo, deemed to have a 'safe' 'ripe' scent. Large game, such as tapir, monkey and peccary, with their 'evil', 'overripe' odours are considered far less edible. The same applies to fruits, which can be classified as either 'ripe' or 'overripe'.[30]

In addition to regulating the selection and processing of foods, olfactory considerations play a central role in determining the combinations in which foods can be eaten by the Desana. For instance, most meats are said to have a male odour, while most fish are deemed to have a female scent. Meat, thus, cannot be cooked or eaten together with fish, for this would be similar to an indiscriminate confusion of the sexes. According to the Desana: 'To mix meat and fish together would be like committing adultery.'[31] There are other sets of rules governing what meats and vegetables or fruits may be consumed together, which kinds of fish may be eaten together, and so on. It is forbidden, for instance, to eat the meat of black brocket deer together with avocados, because the potent generative odours of these foods would render their consumer overly fertile and she or he would engender twins – considered an aberration by the Desana. So important are the complex rules concerning food combinations in Desana culture, that when people are eating, a bystander will sometimes say, not 'eat up', but *mereké*, 'combine well'.[32]

The Batek Negrito of the Malay Peninsula have an equally extensive set of rules regarding the combination of foods. They believe that animals with different odours should not be cooked over the same fire. The odours of some species, however, are said to be so similar that it is acceptable to combine the meats of those animals in food preparation. Bamboo rats and dusky leaf monkeys, for example, are customarily cooked together because of their presumed similarity of odour.[33]

Analogous rules apply to vegetables. Most Batek Negrito vegetables, such as palm cabbage and wild banana flowers, can be cooked together with starchy food, such as rice and tubers. Animals can generally be cooked with starchy food, but not with vegetables (except for small amounts of onions or wild ginger used for seasoning). Certain animals, however, such as macaques, are considered so strong-smelling that they cannot even be cooked on the same fire as starchy food. People who break this food code are said to be punished by the *hala'*, the Batek Negrito deities.[34]

These restrictions on food preparation can be better understood if we examine the Batek Negrito myth concerning the origin of food:

> Once two *hala'* brothers came upon a huge bearcat... Its body was three times as large as an elephant's and it was several hundred feet long. They blowpiped it and managed to kill it. Then they butchered it. They threw pieces of meat in different directions and, as they did, they called out the names of various edible plants and animals. The meat became the [plants and animals] named.[35]

This myth, the basic theme of which can be found in cultures all over the world, explains that all foods originally came from one being, a giant bearcat. The deities create the different edible plants and animals by killing the bearcat, scattering its parts abroad and turning each piece into a different kind of plant or animal. For the Batek Negrito, to cook different plants and animals together would be to try and reverse this original process of food creation and blend separate food categories back into a unitary whole. Such an act would, of course, constitute a rejection of the creative work of the deities and naturally provoke their anger.

As odour is the element of foods which is most prone to intermixing, and as mixed food odours are the means by which the deities learn of an infraction of the food code, the Batek Negrito must exercise great care in their food preparation to prevent the odours of animals and plants which should be kept separate from mingling. This is so much the case that the anthropologist Kirk Endicott reports seeing a Batek Negrito woman 'cooking two minnows, three tiny crabs, and two small shrimps over three separate fires'.[36] Significantly, the bearcat, the mythical source of all foods, is held to have a very strong odour by the Batek Negrito,[37] so that one might imagine that the original bearcat contained within its scent the combined odours of all edible plants and animals, which are now separated.

Among the Kapsiki of Cameroon and Nigeria, the distinctions made between edible and inedible foods are related to the distinctions made between social classes, specifically the class of farmers and traders which constitutes the mainstream of Kapsiki society (known as the *melu*), and the marginalized blacksmith class (or *rerhè*). Blacksmiths, who also serve as diviners and undertakers, are regarded by the *melu* as dirty and foul-smelling. This is particularly due to their role as undertakers, for the odour of a corpse is considered by the mainstream Kapsiki to be the worst and most polluting of odours. 'Our noses turn away from a blacksmith,' the farmers and traders say, with complete disdain for those they consider their social inferiors.[38]

Blacksmiths include in their diet animals such as horses, donkeys, monkeys, felines, carrion birds, turtles and snakes, which mainstream Kapsiki are forbidden by custom to eat. Such animals are not only considered inedible in themselves, but become even more repulsive as sources of food to the *melu* by virtue of their association with the 'foul' blacksmith class. This mainstream abhorrence of blacksmith foods extends even to the dishes the smiths eat from. When visiting someone from another class, therefore, a blacksmith must come provided with his own drinking bowl. In fact, many blacksmiths wear a bowl as a hat for this purpose.[39]

While the animals mentioned above partake of the smiths' polluting stench, according to mainstream Kapsiki notions, those animals which are considered eminently edible, such as cattle, are thought to have no particular smell. Other animals (for example, certain birds) which are not so obviously edible to the mainstream Kapsiki, but still form a part of their diet, are said to be distinguished by their odours of edibility.

The blacksmiths, however, have a different perspective on the matter. They agree that cattle (the epitome of the class of edible animals) have no odour. Where they disagree with the *melu* is in regard to the list of animals considered malodorous and inedible.

The smiths, who do not consider themselves or the corpses they deal with to be bad-smelling or contaminating, do not think that any of the animals they customarily eat – from donkeys to snakes – smell inedible either. Only those animals the smiths won't eat, such as dogs, have the stench of inedibility for the smiths. The conflicting ways in which the *rerhè* and the *melu* categorize the same potential foodstuffs as edible or inedible illustrate just how much what smells like food depends – in the end – not on the food itself, but on the value system of the group doing the smelling.

The Kapsiki case also demonstrates that conceptions of the edible and the inedible, the fragrant and the foul, differ *within* societies, not simply between them. Such differences are indicative of the *interested* character of olfactory classifications. That is, it would clearly be misleading to treat the Kapsiki olfactory code as a unified system of meanings and values that can be understood in its own right, independently of the relations of social domination which that code can be seen as supporting. Rather than being shared, the meaning(s) of Kapsiki olfactory categories are the subject of social struggle, as the blacksmiths try and resist the malodorous identity foisted upon them by the farmers and traders. Olfactory classifications, like other symbolic classifications, can be sites of social conflict as well as cohesion.[40]

SPEAKING OF SMELL: ODOURS IN LANGUAGE

The absence of a true olfactory vocabulary in European languages has long preoccupied Western scientists.[41] Although the human nose is capable of recognizing thousands of different odours, nearly all of our odour categories – sweet, pungent, bitter, and so on – are borrowed from a limited selection of taste terms. Smells are otherwise designated by reference to the things from which they emanate, for example, the smell of coffee, the smell of paint, the smell of grass. It has been suggested that this poverty of olfactory terms is due to the relative unimportance of olfaction in the West. This leads us to wonder how the olfactory vocabularies of those societies which attach relatively more importance to the power of smell than we do in the West compare with the smell lexicons of European languages.

In many non-European languages, the same terms are used for both smell and taste. The odour categories of the Brazilian Bororo, for example, allude to taste as well as smell. This is not

surprising, given that a large part of the sensation of flavour
actually depends on smell. However, odour and taste vocabularies
are not always the same. The Serer Ndut of Senegal, for example,
have four flavour terms: *sen* – sweet, sugary; *kob* – acidic; *sob* –
insipid, cool, such as raw manioc or unseasoned grains; and *hay*
– piquant, hot, including tastes of salt, pepper and bitterness.
Odours, on the other hand, have a more extensive vocabulary.
There are the five odour classes described earlier: *sun* – urinous;
hot – rotten; *hes* – milky or fishy; *pirik* – acidic or acrid (of which
pen, mildly acidic, is a sub-category); and *hen*, fragrant, flowery.
Apart from these, there are also two general categories of *kiili*,
human odours, and *nget*, non-human odours. These terms, with
all of their connotations, can be used to classify and characterize all the different elements that make up the Serer Ndut world.
The donkey, for example, is said to smell non-human (*nget*) and
acidic (*pirik*), and to taste piquant (*hay*).[42]

It is notable that the Serer Ndut sweet taste class (*sen*) is
completely separate from the fragrant smell class (*hen*). This
is different from our term 'sweet', which can mean both sugary
in taste and fragrant in odour. In fact, of course, a 'sweet' smelling
flower may have a very bitter taste, so that our general use of
the term sweet can, at times, be misleading. On the other hand,
while the Serer Ndut take care to distinguish odours from flavours, they do not make many distinctions within these sensory
domains. Thus there is only one taste term (*hay*) for salty, peppery
and bitter, and one odour term (*hen*) for fragrant.[43]

Another African language which merits scrutiny in connection
with the present discussion is that of the Kapsiki of Cameroon.
The Kapsiki have fourteen distinct smell categories, all of which
are in general use.[44] These terms are given in Table 3.

Most of the terms in the Kapsiki smell vocabulary can be
applied to various things considered to have a similar odour. Thus
'urduk'duk, milky-scented, can be applied to white people and to
toads by the Kapsiki, just as English-speakers apply the term
orange not only to the fruit, but to anything considered to have
a similar colour. The most interesting thing about the Kapsiki
lexicon, however, is that there are substantial disagreements as
to what items actually belong in each smell class. This dissension
occurs not only along social class lines (as we saw in the last
section), but also along gender lines. A male member of the
marginalized blacksmith class, for example, will think that the first

Table 3: Olfactory classification system of the Kapsiki of Cameroon

1. *Mèdèke*: the smell of various animals
2. *Vèrevère*: the smell of civet
3. *Rhwazhake*: the smell of urine
4. *'Urduk'duk*: the smell of milk
5. *Shireshire*: the smell of the faeces of various animals
6. *Ndrimin'ye*: the smell of spoilt food
7. *Ndaleke*: the smell of rotting meat or of a corpse
8. *Duf'duf*: the smell of white millet beer (*mpedli*)
9. *Hes'hese*: the smell of roast food
10. *Zede*: the smell of edible food
11. *Kalawuvè*: the smell of human faeces
12. *Kamerhweme*: the smell of old grain
13. *Rhweredlake*: the smell of fresh meat
14. *Dzafe*: a fleeting smell of any kind

odour category, *mèdèke*, applies to snakes, (fresh) fish, pelicans and dogs. A male member of the dominant farmer and trader (*melu*) class, on the other hand, will place in this category snakes, fish, horses, shrews, fresh blood, and the despised blacksmiths. Women of both classes, by contrast, will classify snakes, fish, billy-goats and boars as *mèdèke*. The scent of snakes and fish is thus the characteristic scent of *mèdèke*, agreed on by all groups. The placement of any other odours in the *mèdèke* category, however, depends on the person's class and gender.[45]

The Desana of Colombia, a highly smell-conscious people, have two general odour categories: good smell, *waro sëríge*, and bad smell, *nyéro sëríge*. The paradigmatic good smell for the Desana is that of flowers, which are used extensively in Desana symbolism. Menstrual blood, on the other hand, is considered the epitome of bad smells. The specific odours of different plants, animals, places and so on are designated in the language spoken by the Desana by placing the word odour, *sëríri* or *suri*, after the name of the thing in question. Thus the smell of monkeys, for example, is called *gahkí sëríri*, monkey odour; the smell of anteaters, *bugu sëríri*, and so on. Certain smells can be extended to apply to a number of things. *Pore suri*, flower odour; for instance, can be used generally to mean fragrant, good-smelling, as in *nëngë pore suri*, fragrant forest, or *vai pore suri*, good-smelling fish.[46]

Although Desana taste terms do not serve as odour categories, the Desana olfactory vocabulary is somewhat similar to that of

English in that odour terms, unlike terms of flavour or colour, are usually intrinsically associated with the things to which they refer. Just as we might speak of a smell of roses or a doggy smell, the Desana speak of *vehkë sëríri*, tapir odour. The difference is that English-speakers refer to only a limited range of items in this way, while the Desana speak of virtually everything – from the odour of shamans to the odour of ants – and that each of these odour categories is imbued with ritual and symbolic significance.

It is not only odour which is classified and thus expressed in different ways through language, but also the act of smelling. In English we use but one word to refer to both inhaling and emitting odours – namely, to smell. Other languages are a good deal more explicit. The terms listed in Table 4 for the act of taking in odours come from a seventeenth-century dictionary of Quechua, the language spoken by the Incas and still spoken in the Andes.

Table 4: Inca olfactory terms [47]

1. *Mutquini*	to smell something
2. *Mukacuni*	to smell a good odour
3. *Aznacuni*	to smell a bad odour
4. *Mutqquichacuni*	for a group to smell something together
5. *Mutqquichini*	to make someone smell something
6. *Mucacumuni* or *mutqquimuni*	to secretly sniff out what is being planned
7. *Aznachicun*	to have oneself or let oneself be smelled
8. *Camaycuni*	to come across a food odour, to inhale, to inspire

A similar variety of verbs expresses the emission of odours. The extent to which the Incas took care to distinguish between kinds of smelling indicates something of the attention they paid to the olfactory process and its different effects. For a group to smell something together – for example, incense burning at a ritual – was evidently a significantly different experience for the Incas from that of an individual smelling something alone. Likewise, the experience of smelling a pleasant odour and that of smelling an unpleasant one were considered sufficiently different to merit separate terms.

In conclusion, we find in the languages of other cultures a

greater variety of olfactory terms than is available in English, or indeed any of the other languages of Europe. There is a general tendency, however, for odours (like flavours, but unlike colours) to be classified according to a division of pleasant/unpleasant. This points perhaps to the primordial importance of smell as a means of discriminating between what is safe and enjoyable, and hence pleasing to the human organism, and what is dangerous and hurtful, hence displeasing. There is also a tendency for odour terms to refer to the sources from which odours emanate – as in 'tapir odour' – and not to some essential quality of the odour itself, such as pungency. The reason for this would seem to be the widely perceived intrinsic association of odours with their sources.

In closing, it is important to realize that a limited olfactory vocabulary does not preclude extensive olfactory symbolism. Although there may not be many ways to speak about odours, an immense number of odours can still be recognized, charged with social and emotional content, and remembered. In fact, it may be that odours tend to be processed in a direct, non-verbal way by the brain and so elude expression through language. This means that to understand the role of odour in different cultures, one must go beyond language and explore the realm of practice.

SCENTS OF SELF: ODOUR AND SELF IDENTITY

Odour is an essential means of defining not only different classes in many societies, but also the individual self. Among the Ongee people of the Andaman Islands, for example, the identifying characteristic and life force of all living beings is thought to reside in their smell. The Ongee hold that a person's odour emanates from his or her bones, which themselves consist of condensed smell, just as the odour of a plant or tree originates in its stem or trunk. It is through catching a whiff of oneself, and being able to distinguish that scent from all the other odours that surround one, that one arrives at a sense of one's own identity in Ongee society. This understanding is reflected in the Ongee practice of touching the tip of one's nose when one wishes to refer to oneself as 'me'. For the Ongee, the tip of the nose stands both for one's olfactory organ and one's odour.[48]

The Bororo of Brazil relate personal identity to smell as well. They differ from the Ongee, however, in that the Bororo believe that the characteristic smell of a person is a combination of the odours of body fluids – associated with the life force – and one's breath – associated with the soul, rather than simply an emanation of the bones.[49] The Serer Ndut of Senegal, similarly, say that an individual is animated by two different forces, each defined by scent. One is considered to be physical and is associated with the body and the breath. The other is purely spiritual. This spiritual scent-soul is believed to survive the death of the individual to be reincarnated in a descendant – an example of a transmigration of scent, as it were. According to the Ndut, one can tell which ancestor is being reincarnated in a child by the similarity of the child's scent to that once possessed by the deceased.[50]

Smell plays a central role in greetings and interchanges between individuals in many non-Western cultures, perhaps because of its identification with the essential self in those cultures. Among the Ongee, for example, on greeting someone, one asks not 'How are you?', but *'Konyune? onorange-tanka?'*, 'How is your nose?', or literally, 'When/why/where is the nose to be?' If the person feels 'heavy' with odour, the enquirer politely sniffs some of it away. If, on the other hand, the person feels she or he is low on odour-energy, the enquirer will provide an infusion of extra scent by blowing on her or him.[51]

In India, the traditional method of greeting was to smell someone's head. Thus the Vedas speak of the satisfaction fathers take in smelling the heads of their children after returning from an absence. This act was as meaningful and affectionate as a kiss or hug would be in the West. One Vedic passage proclaims: 'I will smell thee on the head, that is the greatest sign of tender love.'[52]

When odours are thought to convey the essence of a person, olfactory interchanges also contain an element of danger, and care must be taken to regulate the interaction. The Amazonian Desana, for example, manifest the same concern with respect to how human odours are combined as they show with regard to food odours mingling. All the members of a tribal group are said to share a similar odour. Since marriage, according to the Desana, should only take place between persons with different odours, this means that one must always look to other tribal groups for one's spouse. This belief is expressed ritually by one tribal group presenting another with meat and receiving in return a gift of fish, or by the exchange of different-scented ants.[53]

The Batek Negrito of the Malay Peninsula, in turn, say that

the odours of close relatives of opposite sexes should not be allowed to mix. Such mixing is considered to occur primarily during sexual intercourse, but may also transpire whenever two persons sit too close to one another for too long. This mingling of overly similar personal odours is said to result in disease for the parties concerned and for any children they might conceive.[54]

The Temiar people, also of the Malay Peninsula, believe that an 'odour-soul' resides in a person's lower back. If one passes too closely behind a person, this odour-soul is disturbed and mingles into the passer-by's body, causing illness. The Temiar say that this dangerous olfactory mingling can be avoided by reciting 'Odour, odour,' when walking behind a person, so that the odour-soul is prepared for the intrusion into its air space.[55]

These various taboos on olfactory mixing can be seen to have important social functions. The Desana restriction on marrying someone of the same odour, like the Batek Negrito restriction on mingling one's odour with a close relative of the opposite sex, is, effectively, an incest taboo. In addition to discouraging inbreeding, these taboos enlarge the social network of the group by forcing people to marry out. The Temiar taboo against walking behind a person's back without reciting 'Odour, odour' ensures that no one will be taken by surprise by someone silently approaching from behind. The fact that *smell* has been chosen as the medium through which to convey these notions brings out how closely allied odour is with personal identity and individual integrity in these cultures.

To conclude our exploration of the olfactory self, it should be noted that odours are not static in the individual, any more than they are in the environment. In the olfactory typology of the Suya of Brazil, a person's scent is said to alter with age, from the strong odour of childhood to the pungent odour of old age.[56] The Desana say that personal odour, called *oma sëríri*, is a combination of one's natural odour, odours acquired through the food one eats, odours caused by emotions and periodic odours related to fertility. These last smells are thought to resemble the perfume of ripe fruit or of sweet roots and certain aromatic herbs. Their emanation is compared by the Desana to the sudden bursting open of a seed pod, signalling procreative readiness. Young people who are in this olfactory state must take care to seek out partners whose clan odour is distinct from their own, in accordance with Desana marriage rules. Thus a Desana father, seeing his son courting a girl, will utter the same expression customarily spoken at meals: 'combine well!'[57]

Why is odour identified with the self in so many cultures? The answer is no doubt partly rooted in the extremely widespread association between smell, breath and life. Smells are both carried on the breath and taken in by the breath as it provides life-giving air to the body. Body fluids, also commonly associated with the life force, all have distinctive odours as well. These bodily odours, emanating as they do from the interior of a person, give the impression of conveying the person's *essence*, or essential being.

Odours also tend to make a forceful physical and emotional impact on one. Thus, smelling an article of clothing belonging to a person will often give a much stronger impression of that person's presence than seeing the piece of clothing would. Furthermore, scent trails, which we all leave behind us wherever we go (although this fact hardly ever enters our consciousness in the modern West) evidence the particular paths an individual takes in the world. All these factors would appear to inform the ways in which different cultures 'make scents' of the self.

THE ODOUR OF THINGS: (C)OSMOLOGIES

Every osmology considered here forms part of a cosmology, or body of ideas concerning how the universe is ordered. While experienced as personal and local, odours are often conceived of as cosmic in their operation and effects. In what follows, we shall be examining how, in many cultures, the order of things is given in and through the odour of things.

In Batek Negrito religion, for example, the deities (who have dew for blood) live in a cool, fragrant land of fruit blossoms in the sky. When spring comes on earth, the deities are said to throw down a certain quantity of these blossoms so that the trees can flower and bear fruit. Underneath this celestial land is the sun, which is believed to have a bad smell of raw meat. The sun is said to acquire its stench by passing through the putrid land of corpses after it sets in the west. It is sometimes called 'anus of the sky' by the Batek Negrito because its heat rays are perceived as foul emanations dropping sickness on the humans living below.[58] The Batek, who avoid excessive heat themselves, say that the moon runs away from the heat and foul smell of the sun. Therefore, when the sun goes under the earth at night, the moon rises in the sky. The moon, in contrast to the sun, is conceptualized as cool, fragrant and healthful.

Batek olfactory beliefs and practices are based on this cosmic opposition between life-giving fragrance, associated with flowers and coolness, and deadly stench, associated with blood, decaying flesh and heat. Humans, with their hot blood and their liabilty to sickness and death, resemble to some extent the hot, putrid sun. In their preference for coolness and their delight in fragrance, however, they imitate the moon and the deities. In Batek rituals fragrance is used to convey notions of life and to attract, while stench, as a sign of corruption, is either controlled and suppressed or used to repel.

The olfactory principles underlying Batek Negrito cosmology are similar in certain ways to those that animate the cosmos among the Bororo of Brazil. The opposition between fragrance and stench manifests itself among the Bororo, however, as an opposition between two groups of supernatural beings, the *bope* and the *aroe*, each with its own domain of influence. The *bope*, described as dark, hot, repulsive little men, are said to reek of *jerimaga*, a putrid, musky smell.[59] The *aroe*, described as cold, transparent winds, smell of *rukore*, sweetness. The former are identified with animate nature and with fire, rain, the sun and the moon, and the latter with mountains, rocks, lakes, rivers and the ideal essences of all beings. Within the human being, the Bororo associate the putrid *bope* with body fluids and the life force, and the fragrant *aroe* with the breath and the soul.[60] In general, the *bope* stand for organic transformation in Bororo cosmology, while the *aroe* embody timeless form.

The contrast between these two groups of supernatural beings and their odours is mirrored by the two kinds of Bororo shamans: the *bari* and the *aroe etawarare*. The *bari* shaman is associated with the *bope*. He is considered to be 'hot', sexually active and rotten-smelling. The *aroe etawarare* shaman is associated with the *aroe* and is 'cold', sexually controlled and sweet-smelling. The former is a 'master of time' (transformation), and the latter a 'master of space' (form). The primary duty of each shaman is to intercede with his respective patron deities on behalf of the Bororo people.[61]

For the Bororo, therefore, odours are representative of the basic cosmic forces of life and spirit, change and structure. The olfactory manifestations of the natural world signal the operation of these forces and remind the Bororo of the practices they must observe in order to maintain a viable place for themselves, their ancestors and their children in the cosmic order.

In the culture of the Ongee people of the Andaman Islands, the cosmos is also ordered by scent. Indeed, smell, for the Ongee, is *the* vivifying principle of the organic world, much as among the Bororo. In striking contrast to the Bororo osmology, however, the beings and forces of the inorganic spirit world are conceptualized as *inodorate* by the Ongee. Spirits, called *tomya* and considered to be the souls of dead humans, are likened to winds, and their home is said to be in the sky. Unfortunately for living humans, however, *tomya* are avid for the smell of life, which they will steal from humans whenever they have the opportunity. The consequence of having one's life-smell taken is to die and become a spirit oneself.[62]

The role of the *tomya* in human existence is not all negative, however. In fact, spirits are necessary for the very continuance of human life since, according to the Ongee, it is only when a spirit is absorbed by a fertile woman that a new human being is created. The spirits, for their part, depend for their existence not only on the odour provided them by humans, but on the ritual offerings the Ongee make to them.[63]

The concern of the Ongee is to assure that a proper equilibrium and interchange is always maintained between odorate humans and inodorate spirits so that the cosmos remains both stable and dynamic. This interchange is likened by the Ongee to the act of breathing itself, during which one is continually taking in and releasing odour. As their numbers decrease to less than a hundred, however, the Ongee are discouraged about their ability to maintain this cosmic order. In the words of an Ongee man:

After our ancestors started getting too many things from outsiders and started going to Abardeen (Port Blair) we had many wars – many died ... Now we are so few left. If we do not [continue our ritual olfactory exchange with the spirits] then all of us will keep on dying there will be no more Ongee. Just *tomya* and *tomya* – nobody to marry no give and take nobody will be here only *tomya* and outsiders. Then *tomya* too will go away forever because outsiders do not give to *tomya* as we do to them. The *tomya* too will die and only outsiders will live.[64]

COLOUR, SOUND AND SCENT: LINKING THE SENSES

Osmologies, as we have seen, are integrated into the social and cosmic orders of the societies which employ them. At the same time, osmologies are related to other schemes of sensory symbolism: odour-meanings are linked to colour-meanings and sound-meanings, and so on. In the culture of the Dogon people of Mali, for example, odour and sound are believed to be fundamentally alike because they both travel on the air. The Dogon will therefore speak of 'hearing' a smell. Furthermore, speech is said to have an odour. Good speech is said to smell fragrant – specifically, to have the odour of oil and cooking, a scent highly valued by the Dogon. By contrast, nasal speech, which is associated with witches, is thought to smell of decay, for it sounds stagnant, as if caught between the nose and the throat. As the Dogon perceive the functions of speaking and smelling to be interdependent, a girl who at the age of ten or twelve still makes mistakes of grammar or pronunciation will have her nose pierced as a corrective. Likewise, making sure one's breath smells sweet is thought to improve the quality or content of one's speech.[65] The meaning of smell, consequently, is intrinsically related to that of sound in Dogon culture.

In certain cultures such sensory correspondences are elaborated into a comprehensive system of interrelated sensory codes. In traditional Chinese thought, for example, odours correspond to flavours, and flavours correspond to colours, which in turn correspond to musical tones, and so on. Thus, a goat smell is associated with a sour taste, the colour green and the musical tone *chio*, while a fragrant smell is associated with a sweet taste, the colour yellow and the musical tone *kung*, as can be seen in Table 5.

Table 5: Chinese table of correspondences

Element	Odour	Taste	Colour	Tone	Season	Direction
Wood	goat	sour	green	*chio*	spring	east
Fire	burnt	bitter	red	*chih*	summer	south
Earth	fragrant	sweet	yellow	*kung*	–	centre
Metal	rank	acrid	white	*shang*	autumn	west
Water	rotten	salt	black	*yu*	winter	north

The basis of this system of correspondences is the 'Theory of the Five Elements', which holds that the cosmos is composed of water, wood, fire, earth and metal. These elements are joined in what is called the 'mutual production order'. This order, which is the governing pattern of the universe, unfolds as follows: water produces wood (by analogy to the way water enters into the substance of plants), wood produces fire (by burning), fire generates earth (in the form of ashes), earth produces metal (by yielding ores), and finally metal produces water (as when metal liquefies in the smelting process).[66]

As the preceding account suggests, the so-called elements are not so much fundamental substances as they are fundamental processes, transforming into each other in a never-ending cycle. Apart from the correspondences given in the table, to every element there also corresponds a planet, kind of weather, style of government (enlightened, relaxed, cautious), kind of grain, wild animal, domestic animal, body orifice, sensory organ, emotional state, and so on, virtually without end.

In addition to providing a comprehensive theory of how the universe is ordered, the system of correspondences helped regulate human behaviour in traditional Chinese society. For the peasantry, the system served as a kind of farmer's almanac, predicting weather conditions and instructing people in how they should tend fields and animals at different times of the year. For the Emperor and those attached to the courts, the system served both as a guide to statecraft and a book of etiquette, specifying what activities and rituals should be performed (or not performed) in order to stay in power.

The proscriptions and prescriptions weighing on the Emperor were very rigorous, as the following account of the proper living quarters, mode of transportation, apparel, and diet for the first month of spring attests:

> The Son of Heaven shall live in the apartment to the left of the Green Bright Hall. He shall ride in a belled chariot driven by dark green dragon [horses], and bearing green flags. He shall wear green clothes with green jade. He shall eat wheat and mutton.[67]

Showing proper respect for these ritual forms was believed to exert a beneficial influence over both the natural and social orders, while contravening them was to invite disaster: drought, failed crops, rebellion.

Of the many cultures we have looked at in this chapter, the

one with the most complex system of sensory correspondences is that of the Desana. The Desana believe that every sensory property embodies both a cosmic energy and a moral value. The orginal source of most sensory properties and of life is held to be the sun. The heat and light of the sun are said to produce three primary colour 'energies': white, signifying vital force; yellow, signifying male generative power; and red, signifying female fecundity. These colour energies are held to combine with heat to produce odour, which quality, in turn, gives rise to flavour.[68]

The moon, according to the Desana, produces a complementary set of energies. These are, first of all, blackness and coldness, from which come the three colour energies of light green, light yellow and dark red or purple. Light green is related to young plant growth, light yellow to ripeness, and purple to rottenness. These colours, in turn, are linked to the odours and flavours associated with their symbolic meanings. Thus purple, associated with rottenness, for example, is linked to a putrid smell and an acid flavour – qualities thought by the Desana to be characteristic of overripe fruit and decaying meat.[69] The Desana associate the colour blue with the Milky Way and with the realm of visions and thought. It is the colour of communication between the natural and supernatural spheres and, as such, is connected with the media of such communication: tobacco smoke and incense.[70]

Sounds are divided by the Desana into three basic types: whistles (sustained sounds), buzzes (vibrating sounds) and rattles (percussive sounds). Whistles – as produced, for example, by panpipes and deer – are said to signify an invitation. Buzzes, such as produced by certain flutes, hummingbirds and bees, signify a warning. Rattling sounds, such as produced by drums and woodpeckers, indicate union.

Furthermore, sounds are associated with colours, temperatures, odours and flavours. For example, the whistling sound of panpipes is said to have a red colour, a hot temperature, a male odour and the flavour of a certain fleshy fruit, all of which qualities carry corresponding moral values. Shapes, patterns, spatial location, movements and textures are also included in this system of correspondences.[71]

Virtually everything in the Desana world is coded according to its perceived sensory qualities. An animal, for example, will be characterized not only by odour for the Desana, but also by colour, flavour, voice and other qualities. Together all of these

sensory attributes and their associated values make up the meaning that animal will have in Desana culture. This classificatory
system applies to anything with sensory properties, from a bird
to a house to a dance.

The Desana believe, significantly, that sensory properties are
not only, or even primarily, manifest in the world, but exist within
the human brain. According to the Desana, the brain is made up
of compartments which contain a variety of colours, odours, flavours, sounds, textures and other qualities, all of which have
related moral concepts and which together produce a state of
consciousness. The purpose of the material world, in fact, is
thought by the Desana to be to serve as a reminder of these
ideal sensory and social values stored within the brain.[72]

Thus odours and their classification, while forming a coherent
symbolic system in their own right, are at the same time interrelated with a whole series of other sensory metaphors in Desana
culture and, indeed, to a greater or lesser extent, in all cultures.
The purpose of the present chapter has been specifically to examine olfactory symbol systems. It must not be forgotten, however,
that these exist and acquire meaning within a multisensory
context.[73]

Chapter 4

The rites of smell

In the previous chapter we saw how people in different cultures employ olfactory constructs to help make sense of the world and their place within it. In what follows we shall explore how these olfactory codes are translated into practice through being enacted in ritual contexts.

Rites which involve odours and olfactory symbolism are elaborated around a wide range of activities and events, including courtship, communication with the spirits, funerals, raising crops, and healing – to name a few. By comparing the ways in which smell is ritually employed or evoked in connection with these activities, we will be able to discover some of the characteristics attributed to odours across cultures, and how these characteristics contribute to the meaning of ritual events.

Odours, for example, are often used in rituals for their beauty. This beauty is believed to have the power both to please and attract those to whom the scent is directed – the deities, the guests at a wedding, a lover, and so on. An integrative power is also usually attributed to smell, making scent an excellent means of uniting the participants in a ritual, who all breathe in and are enveloped by the same aroma. The boundary-crossing nature of smell, in turn, is often made use of to help the participants in a rite of passage – for example, a funeral – cross over from one stage to the next: they are symbolically wafted along with the olfactory flow.

At the same time, odours can repel. Odours considered foul may be ritually employed to ward off unwanted presences, such as evil spirits, or ritually controlled in order to prevent such odours from disturbing the social or cosmic order. The smells of childbirth are an example of an olfactory phenomenon deemed to be repellent and disruptive in many societies. These smells must accordingly either be dispelled by the use of other scents, or controlled by secluding both the newborn and its mother.

These and other cultural attributes of odour will be revealed as we explore the different domains of existence in which odours serve to express and create meaning for people. We are left with a sense of the centrality of smell in many societies, and with an appreciation of the variety and complexity of olfactory customs around the world.

A LAUGH OF JASMINE: THE AESTHETICS OF SMELL

A South Indian folktale tells of a king whose laugh would spontaneously spread the fragrance of jasmine for miles around.[1] This poetic image conveys something of the exhilarating pleasure of smell, the pure delight associated with fragrant scents. Fragrance can indeed be a powerful aesthetic experience, expressing an intimately perceived ideal of beauty and grace. Yet what is experienced as fragrant varies greatly from one culture to another. The cattle-raising Dassanetch of Ethiopia, for example, find nothing more attractive than the odour of cattle, a scent which carries notions of fertility and social status for them. Dassanetch men, consequently, decorate themselves with cattle bones and hides and even wash their hands with cattle urine, as well as smearing their bodies with cow manure. Dassanetch women rub their heads, shoulders, and breasts with butter in order to advertise their fertility and make themselves attractive to the men by their scent.[2]

For the Dogon of Mali, the loveliest fragrance is that of the onion. Young Dogon men and women will fry the onion plant in butter and rub it all over their bodies as a perfume. By contrast, among the Tamils of India, it is said that 'you can infuse an onion with many fragrances, but it will never lose its stench'.[3] There, sandalwood and aloewood are favourite scents. Different again are the olfactory preferences of the African Bushmen: they have elaborated numerous folktales around the sweet, seductive scent of rain, to which, they say, no other fragrance can compare.[4]

Not only do concepts of fragrance differ among cultures, so do the modes of employing it. In the Trobriands, mint is boiled in coconut oil while a spell is pronounced over it in order to create a love charm. The magic-maker does not anoint himself with this potion, however; rather he seeks to spill some on the breast of the desired one as she sleeps. This act is supposed to inspire erotic dreams of the magic-maker in the woman, so that when she awakes she will be unable to control her passion for him.[5]

The inhabitants of the South Pacific island of Nauru, for their part, believe that by *drinking* a perfume potion of scented coconut milk, one's body and breath will become irresistibly fragrant. Using carefully guarded formulas, Nauruans combine fragrant flowers with coconut milk to create an aromatic drink which, once drunk, will perfume their bodies over a number of days. Such perfume potions are said to be more effective than regular perfumes as the scent is emitted from inside the body, and not simply the skin's surface.[6]

Another perfumery technique employed by the Nauruans is that of steaming one's body with scent. In this method a little oven is built in the ground and scented flowers and leaves are placed on hot stones within. A Nauran woman will stand over the oven enclosed in a mat to let the fragrant steam penetrate her body completely. Even her head will be covered with the mat in order that she may breathe in the fragrance deeply. This practice is believed to cleanse the body of any unpleasant odours at the same time as it invests it with a lasting perfume.[7]

Perfumes can also be used in conjunction with visual decorations. Sweet-smelling flowers and leaves, for instance, are a common adornment in many cultures. Many different peoples, as well, decorate their faces and bodies with scented paints. The inhabitants of the Trobriands mix charcoal and aromatic chips of wood in coconut oil in order to create a fragrant black paint which they use for tracing lines on their faces.[8] The face paints of the Colombian Desana are made of a combination of aromatic saps, flowers, fruits and pigments. The fragrant yellow sap of a certain local tree, for instance, is used to paint round spots on the faces of Desana men. These perfumed yellow dots represent the fertilizing power of the sun.[9]

Fragrance, in many cultures, is not only a matter of using perfumes to augment one's personal attraction, however. It is also a means of enhancing the aromas of one's food, possessions and living space. Some of the most elaborate examples of such uses of fragrance are found in the cultures of the Near and Far East. Let us look, for instance, at the aesthetics of smell in the United Arab Emirates. While many different perfumes are used in the UAE, the most important and popular are the following: aloewood, ambergris, saffron, musk, rose, jasmine, Arabian jasmine, narcissus, sandalwood, henna and civet. These aromatics are used primarily in the form of oils, which are thought to hold fragrance better and be more aesthetically pleasing than other perfume media. In addition, certain aromatic seeds and leaves are ground into powders and used in combination with the above oils. Only the wealthy can afford to indulge in all of these scents, however, for the price of some of them is quite high. The favoured aloewood, for example, can cost up to $250 an ounce.[10]

The basic palette of scents named above is employed by Arab women to compose their own blends of perfume which will be applied to different parts of the body. Musk, rose and saffron are rubbed on the whole body. The hair is perfumed and oiled with a blend of walnut or sesame seed oil with ambergris or jasmine. The ears are anointed and coloured with a red mixture called *mkhammariyah*, composed of aloewood, saffron, rose, musk and civet. For the neck, aloewood, ambergris, rose, narcissus or musk are used, for the armpits ambergris or sandalwood, and on the nostrils aloewood. Before any perfumes are applied, however, the body must be washed and clean, not only for aesthetic reasons, but also because it is believed that applying scents to an unwashed body invites disease.[11]

The purpose behind using perfumes in Arabia, therefore, is not to mask unpleasant body odours. Rather, their use is aimed at making the body agreeably fragrant. Women of the United Arab Emirates, expressing the importance of fragrance in their culture, say 'We must use lots of smell.' This use of strong scents by women is not, however, designed to make their presence and allure perceptible at a distance. Women, in fact, are not supposed to be perfumed when in the company of men, for in this situation a perfumed woman is said to be like an adulteress. A woman uses perfumes, therefore, only in the company of other women or of her husband and close family. Perfumes pertain to the private realm, rather than the public.[12]

Men in the United Arab Emirates have a keen appreciation of the perfumes used by women. A husband talking of the charms of his wife, for instance, will stress her good smells, as in the following statement:

Her body is rubbed with a paste of [scented oil] and rose

petals; her sleeping gown is redolent of beautiful aloewood. Her hair smells of ambergris and saffron oils. We men like all scents used but have a preference for musk, ambergris, aloewood, and saffron.[13]

Although fragrance is considered to be primarily the domain of women, men may also wear perfumes. The customary scents for men are rose and aloewood which are placed behind the ears, on the nostrils, on the beard and in the palms of the hands. On special occasions children of both sexes will be perfumed as well.[14]

Arab women make their clothes fragrant by perfuming them with both scented oils and incense. Most of the former are used only on dark clothes because they would stain light-coloured cloth. Women's clothes are censed with a mixture of aloewood, musk, ambergris, gum arabic, rose and sugar, while men's clothes (which are usually white and would be darkened by the above incense) are censed with aloewood alone. A teepee-shaped rack is placed over an incense burner for this purpose. After clothes have been washed and dried, they are placed on the rack until it is completely covered and all the fragrant smoke is absorbed by the material. While the smell remains on the clothing for at least three washings, women will still frequently incense clothes after each wash to intensify their fragrance.[15]

Within the house, all rooms, except for the bathroom and kitchen, are censed every Thursday with frankincense, and on festive occasions with aloewood. The bathroom and kitchen are excluded from this perfuming because, even when clean, they are held to be irremediably polluted by bad smells: those of body waste in the case of the bathroom, and those of animal blood and food waste in that of the kitchen. To perfume these rooms therefore would only contaminate the scent of the perfume without eradicating any foul odours.[16]

Although the kitchen is considered a site of bad smells, the aromas of foods are none the less highly prized in the Arab world. Foods are usually covered when they are cooked so as to preserve their aromas, and cooked foods are always served hot so that their aromatic vapours may be appreciated. Spices, such as pepper, anise, cinnamon, clove, ginger and garlic, along with dried limes, are used to add flavour to salty dishes. Saffron, cardamom and rose-water serve this role with sweet dishes. Even beverages will be made more fragrant through the addition of

scents. Cardamom is added to coffee, for instance, saffron may
be added to tea, and saffron and cardamom oil to milk. Drinking-
water, in turn, is traditionally perfumed with frankincense smoke.
The method that is used is to fill a water pot with thick frankincense fumes and then pour in water and cover the pot. (This
practice has the added benefit of helping to disinfect the water.)
It is said that this culinary use of aromatics 'beautifies the food'.[17]

When guests participate in a meal in the UAE, the visit ends
with a round of perfumes. A meal is not considered complete
unless there is this offering to the nose, as it were. As men and
women visit and eat separately, the round of perfumes will either
be offered by a male host to all male guests or by a female host
to all female guests. A nineteenth-century English traveller in
Saudi Arabia describes this rite among men as follows:

> [After] meals, or even at the conclusion of a simple coffee-
> drinking visit, appears a small square box [for burning
> incense] . . . [It] is filled with charcoal or live embers of Ithel
> and on these are laid three or four small bits of sweet- scented
> wood . . . Everyone now takes in turn the burning vase, passes
> it under his beard . . . next lifts up one after another the corners
> of his head-gear or kerchief, to catch therein an abiding per-
> fume, though at the risk of burning his ears if he be a new
> hand at the business, like myself; and lastly, though not always,
> opens the breast of his shirt too, to give his inner man a whiff
> of sweet-smelling remembrance.[18]

In the twentieth century, an Arab woman's perfume box holds
four to eight bottles of perfume, along with glass applicators,
surrounding a container with various sorts of incense. This box is
brought out after coffee has been served and the food trays
removed. First the bottles of perfume are passed around by the
women who are seated in a circle on the floor. Each woman uses
an applicator to anoint herself with scent, the part anointed –
ears, neck, veil, cloak – depending on the kind of perfume. After
everyone has had the chance to generously perfume herself, a
mixture of incense is placed on the glowing embers of a censer.
When the smoke is dense the hostess hands the censer to the
woman seated to her right. She introduces the censer under her
veils to perfume her hair and face, then lets the fragrant fumes
penetrate all of her dress and body, wrapping herself up in her
cloak to better contain the smoke. When she has finished, the

censer is passed to the woman on her right, until everyone has had the opportunity to cense herself.[19]

Bringing in the perfume box is a sign that the visit has come to an end, and when the perfuming ritual is over the guests say their farewells and depart. Guests, therefore, arrive wearing their best perfumes in order to honour the hostess, and leave re-scented with the best perfumes of the hostess, showing that they, in turn, have been honoured. Indeed, the odour of the perfumes will linger for hours after a visit, prompting admiring comments from others: 'You must have been somewhere. You smell nice. Where were you?' The higher the quality of the scents, the more the absent hostess is praised. The social prestige of a hostess, therefore, is enhanced and expanded by the smells she imparts to her guests.[20]

These examples of aesthetic uses of smell indicate both the pleasure many people take in fragrance and its importance in establishing social ties. In the case of the perfume-sharing ritual described above, the participants, who all manifest different scents at the start, at the end are bound together by the same scent, thereby establishing their essential group unity. At the same time the reputation of the hostess is carried by her perfumes so that her social network spreads along trails of scent.

The diversity of perfumery techniques considered here is strik-ing. Fragrance, in the cultures we have studied, is never simply a matter of dabbing oneself with a sweet-smelling liquid out of a bottle. Rather, fragrance is at the centre of sophisticated rites designed to make full use of the aesthetic and attractive forces of perfume within the bounds of social norms. It is significant, as well, that many such traditional perfumery techniques persist in spite of the availability of modern Western perfumes. The Nauru-ans, for example, are avid users of commercially prepared per-fumes; however, when they want a truly effective scent, they rely on their own home-made perfume potions and steams.[21] The women of the United Arab Emirates, for their part, concede that Western spray-style perfumes are more convenient than their own oils and incenses, yet find that they 'smell less beautiful'.[22] Nat-urally, Western perfumes are also totally incapable of supplying the fragrances so meaningful and essential to such non-Western peoples as the Dassanetch, with their passion for bovine odour, or the Dogon, with their love of the aroma of grilled onion, or the Bushmen, who prize the smell of rain.

CIGARETTES FOR THE GODS: COMMUNICATING WITH THE SPIRITS THROUGH SMELL

Fragrant and foul odours, as we have seen in the previous chapters, are not simply matters of aesthetic preference, but are imbued with moral and religious associations. Fragrance is often closely identified with beneficial deities and forces, and foulness, with harmful deities and forces. The Muslim inhabitants of the United Arab Emirates, for instance, say that: 'A dirty, smelly body is vulnerable to evil, the scented person is surrounded by angels.'[23] The olfactory rites of the UAE reflect this belief. The most efficacious scent for attracting angels and dispelling evil spirits is thought to be frankincense smoke. For this reason, children (thought to be especially vulnerable to evil), houses and mosques are weekly censed with frankincense in the UAE.[24]

Similar beliefs are found in other Arabic cultures. In Morocco, for example, foul odours are closely associated with evil spirits. Thus dung heaps are held to be full of evil jinn and considered places of spiritual danger. Furthermore, it is traditionally held that breaking wind inside a mosque will blind, or even kill, the angels therein. Even outside a mosque, the action is so closely associated with harmful spirits, that the spot where it occurred may be marked by a small pile of stones, as if to trap the evil jinn inside. On seeing the cairn of stones, passers-by will spit on it or throw another stone on the pile. So strong is the taboo against breaking wind in public that, among the Berber tribes of Morocco, people have reputedly committed suicide in consequence of such an act. Fragrant scents, of incense or rose or orange water, in contrast, purify the body of evil influences and ally one with the forces of good.[25]

In such cases it is evident that maintaining oneself in good odour is not simply a means of being personally attractive or socially nice, but a matter of sanctity and sin, life and death. Likewise, olfactory rites take place not on the periphery of religious life, but at the heart of the sacred, expressing and enacting a culture's greatest hopes and fears – desires for divine sanction and aid, dread of divine rejection and injury, concerns over social and bodily boundaries, and so on.

Sacred rites of smell are common to many peoples. The aromatic shrine can be found the world over, offering up scents for the pleasure of the gods. In Mexico, the Tzotzil people dedicate

to their deities candles and copal incense, which they refer to as
'cigarettes for the gods'. Hindu temples are redolent with odours
of sandalwood and other aromatics. The Nigerian Songhay pour
out perfumes on the altars of their gods. The Dakota of the
Western Plains send up smoke signals of burning sweet grass to
their deities.[26] These olfactory rites are similar on one level, yet
each has a unique meaning within the context of the particular
culture which practises it.

The Chewong, an aboriginal people of the Malay Peninsula,
for example, consider odour the fundamental means of interaction
with the spirits. Chewong children wear a piece of wild ginger
tied to a string around their necks in order to keep harmful
spirits away by the pungent smell. Good spirits, in contrast, are
attracted by and 'fed' with an incense of fragrant wood, ritually
offered to them every night. During this rite the Chewong shaman
takes some of the incense smoke in his fist, puts his fist to his
mouth and blows in four directions, after which he prays to the
spirits for divine protection. The smoke is believed to carry
the shaman's words up to the spirit world. On no account must
this incense offering be neglected as this would interrupt the
communication with the spirits and endanger the Chewong community.[27]

One class of female spirits, called 'leaf-people', is thought to
be especially fond of aromatics. These spirits are said to live in
flowers and leaves. They wear garlands of sweet-smelling leaves
and flowers in their hair and have bundles of other sweet-smelling
leaves in their loincloths. The leaf-people serve as spirit guides
to Chewong shamans, teaching them magical songs. When the
shamans hold a ritual séance to sing these songs, the leaf-people
are said to come and sit and swing on the strings of plaited leaves
made ready for them. When many leaf-people come to a séance,
the meeting-house is believed to be filled with their fragrance
– the only aspect of the spirits perceptible to non-shamans. This
scent is considered so beautiful by the Chewong that the women
often cry out of sheer emotion when they smell it.[28]

The Chewong's neighbours, the Batek Negrito, also believe that
the spirits are very sensitive to smell. The Batek Negrito spirits,
the *hala'*, are said to live in a perpetually fragrant land of fruit
blossoms. The *hala'* are exceedingly fond of the scents of flowers
and incense and strongly dislike those of blood and burning hair.
Batek olfactory rites, therefore, are centred on either propitiating

the *hala'* with the former or taking care not to offend their sensibilities with the latter. When a Batek Negrito has broken a taboo, for example, she must cut herself and offer up some of her blood as a guilt-offering. In order for the spirits not to be left with the unpleasant odour of blood in their nostrils, however, immediately afterwards the guilty party will wipe the blood off the wound with a fragrant leaf and then burn the leaf as an incense. Likewise, if a thunderstorm threatens a Batek Negrito camp, the Batek will burn wild ginger or aromatic gums in order to influence the *hala'* to stop the storm with the sweet scent. If the storm lingers, the Batek resort to a different strategy: they burn some human hair, in the hope that the stench will convince the spirits to stop the storm where fragrance could not.[29]

In certain cultures, odours not only help one communicate with the spirits, but make it possible for one to temporarily *become* a spirit. In the Afro-Brazilian spirit possession cult, Batuque, for example, incense sets the stage for the arrival of the spirits:

> A young woman enters the pavillion swinging a small charcoal brazier by its long wire handle. Incense is scattered over the smoldering charcoal, yielding a thick fragrant smoke. It is believed that this smoke has cleansing or purifying properties and, as the brazier is carried around . . . some members of the audience stand up and lean over to get a puff of the smoke on their bodies, crisscrossing their extended arms over the brazier as they do so.[30]

This act of censing serves the purposes of purifying the ritual space and participants, thus making them fit for the presence of the spirits, and of helping induce a trance-like state in the participants. The entrance into a trance state is further aided by intense rhythmic drumming and dancing.

The incense burnt depends on the particular spirits summoned. One group of possessing spirits, known as *jurema*, for instance, regard the type of tree called *jurema* as sacred. In a Batuque ceremony dedicated to these spirits, therefore, the leaves of this tree will be burnt as incense. *Jurema* leaves, bark and flowers will also be made into a tea, which is then placed in a bowl under the altar for the spirits to drink. As soon as the participants in the ceremony smell the smoke of the *jurema* leaves, they are prepared for the arrival of the *jurema* spirits. Once in possession of the bodies of their followers, the spirits will offer advice to those who seek it and perform rites of curing.[31]

In the *zar* spirit possession cult of Northern Sudan, odours are similarly fundamental. Breathing in foul odours is thought to render one vulnerable to being invaded by a spirit, while the spirits themselves are attracted to individuals by fragrance and often demand perfumes as payment from their hosts. (In fact, persons possessed by spirits will sometimes drink perfume or ask for perfumed cigarettes.) Censing the possessed person is thought to encourage the spirits to make themselves and their demands known.[32]

One woman describes the olfactory sequence of a spirit possession as follows:

> I was at a wedding [a highly perfumed event] and dancing . . . One Sudani [stranger] man came close to me . . . But he stank so terribly from sweat that I fainted immediately.
>
> My family brought me home and censed and perfumed me . . . I dreamed this song:
>
> What needs have I?
> We want henna, incense,
> A bottle of perfume on which there appears the face of a man . . . [33]

In this case the invasive spirit is attracted by the scents and finery of the wedding, enters a woman who has been rendered vulnerable by breathing a foul odour, and then makes its demand for perfume known after being censed. At the same time male foulness (the sweat of the Sudani man) is transformed into male fragrance (the bottle of perfume with a man's face) through the intercession of the spirit.

During *zar* rituals spirits are convoked en masse through censing and drumming to enter the bodies of their, usually female, hosts. These possessing spirits are thought to be a mischievous kind of jinn who manifest themselves as a number of foreign stereotypes: Ethiopian prostitutes desiring wedding finery, Arab nomads requesting rancid butter to smear on their hair and sour camel's milk, or, most powerful of all, Western Christians who ask for khaki suits, walking canes and alcohol. The spirits convoked by scent thus are also representations of human foreigners, and through the manifestation of the former, a critique and appropriation of the perceived characteristic traits of the latter is accomplished.[34]

An interesting variation on such uses of smell as a medium for the supernatural is provided by the Ongee of the Andaman Islands. The Ongee, who equate smell with life, as we learned in the last chapter, believe that the spirits are inodorate and that they hunt out the living in order to steal their life-smell. Ongee olfactory rites, therefore, are aimed not at attracting the spirits by smell, but at reducing smell in order to keep the spirits away. Various techniques are used in order to accomplish this. Applying clay paint to the body, for example, is considered an effective means of restricting the release of telltale body odours. Smoke, in turn, is used abundantly by the Ongee to mask their odour. When moving from one place to another the Ongee will walk in single file, stepping over the tracks of the individuals in front of them. This procedure is supposed to mix the odours of individuals together, making it difficult for the spirits to track down any one person.[35]

There are times, however, when the Ongee wish to communicate with the spirits, not simply avoid contact with them. A common way for the Ongee to establish such communication is by painting their skin with a clay paint design. The pattern of painted and unpainted skin is said to affect the quality of the smell released by the body. The patterned odour thus emitted is thought to carry a coded message to one's ancestral spirits, who will help protect one from the general run of smell-hungry spirits.[36]

Besides, the Ongee shaman periodically visits the spirit world to gather information from the spirits. In order to do so he must temporarily become like the spirits, which entails losing his odour. This is accomplished by letting the spirits suck up his smell and then carry him up to their world in the sky. While with the spirits, the shaman gains important knowledge about the locations of food resources and the future movements of the spirits. In return he offers the spirits tools for cutting, which they lack, and pledges on behalf of the Ongee to make certain foods available to them. The spirits thus agree to let the shaman return to the world of the living.[37]

These are but a few examples of the ways in which interaction between the human and divine spheres is ritually governed by odour in different cultures. In the sections which follow, we shall be introduced to many further instances of olfactory communication with the spirits.

SANDALWOOD, SUGAR AND CROCODILE CLAWS: OLFACTORY RITES OF PASSAGE

Rites of passage – that is, rituals which transport a person from one condition or social status to another, such as the rites of birth, puberty, marriage or death – are often occasions of olfactory symbolism.[38] Babies, for example, come into the world in a gush of natural smells and need to be socialized into the odours of culture according to some customs. In order to accomplish this, the inhabitants of the Tanimbar Islands of Indonesia traditionally 'smoked' a newborn child over the household fire for the first few weeks of life.[39] In many cultures, in fact, the odours of childbirth are considered a threat to health. In Northern Sudan, a bowl of Nile water and perfumes protects visitors to a new mother from the 'impure' odours of birth.[40] Among the Bororo of Brazil, such odours are thought to afflict those who smell them with listlessness. The midwife therefore buries the afterbirth and burns any items, such as clothes or mats, which have come in contact with birth fluids. Immediately after a child is born, the father must take a thorough bath in order to cleanse himself of the smell of birth. In fact, it is only when a child receives a name, or, in other words, becomes a socialized human being, at the age of about six months, that she or he ceases to be a serious olfactory liability.[41]

Odour symbolism surrounding puberty, similarly, typically centres on a ritualized olfactory passage from nature to culture. However, in this case, the symbolism usually differs according to the sex of the ritual subjects. Among the Suya of Brazil, boys are supposed to lose the strong asocial smell of childhood and attain the ideal bland smell of adult men after undergoing the male puberty rituals. This olfactory transformation is emphasized by the bland smell of the nut oil with which adolescent boys paint themselves. Girls, however, are believed to become even more strong-smelling and asocial as they reach puberty, for their incipient fertility allies them with the forces of nature.[42]

Among the Colombian Desana, a girl, on the occasion of her first menstruation, is secluded in a small room and visited three times a day by the shaman, who blows tobacco smoke from his cigar over her. The sacred and purifying odours of the shaman's tobacco smoke help counteract the 'evil' odours of menstruation. For the Desana, the odour of menstrual blood is the most disgusting and polluting of all odours, and menstruating women are likened to wild animals, outside of all cultural norms. While menstruation makes a girl a 'marriageable woman', therefore, it also renders her a 'dangerous animal', who must be carefully controlled during her menstrual periods. The tobacco smoke blown over a girl on the occasion of her first menstruation enables her to emerge from her outcast condition as an acceptable member of the community.[43]

The odours of menstruation are also considered harmful among the Hua of the Highlands of Papua New Guinea, at least as far as men are concerned. Male initiates, for example, must avoid eating foods said to smell of *be'ftu*, menstruation. This class of foods includes the meat of certain species of possum, a kind of mushroom and two kinds of yam. The Hua believe that to partake of any of these foods would cause the youth to degenerate physically.[44]

The male undergoing initiation must also avoid inhaling when in the vicinity of a menstruating woman, and he must refrain from sexual relations so as not to become 'polluted' with feminine emanations. The reason for this extensive list of taboos is that the Hua define gender more in terms of fluids and scents than in terms of external anatomy. The interior of a girl's body at menarche is conceived of as being dark, juicy and smelling of decay, in contrast to the adolescent male body, which is supposed to be white, hard and odourless on the inside. As these gender-determining substances are considered to be transferable, changes of gender are not uncommon among the Hua:

> a genitally male person may be classified as female through his contamination by female substances, and a genitally female person may be classified as male through the transfer of pollution out of her body ... [In short, a] person's gender does not lie locked in his or her genitals but can flow and change with contact as substances seep into and out of his or her body.[45]

The flow of scents and fluids must be strictly regulated during the period of male initiation because the purpose of the male initiation rites is, of course, to instil a masculine identity in the youth – a difficult task among the Hua, where gender identity is so changeable.

Menstrual odours are, in fact, considered polluting in many cultures around the world.[46] Among the Amazonian Bororo, a menstruating woman is thought to contaminate all food or water she contacts with her stench.[47] In the United Arab Emirates, as in other Arabic cultures, a menstruating women is considered so irredeemably foul that she cannot use any perfumes until after her period is over and she has taken a ritual bath.[48] Such beliefs are often based on a conception of menstrual fluid as a kind of putrid blood, combining the danger associated with blood with that associated with excrement. Furthermore, each menstrual period has the negative connotation of being a failed birth.

This antipathy for menstrual odours is not universal, however. The Ethiopian Dassanetch, for instance, say that menstrual blood has no significant odour. Menstruation is called 'the rain of a woman', and is positively valued in that, like rain, it is held to promote fertility. Older, infertile women, however, are said to stink of fish, for they, like fish, are placed outside the beneficial cycles of nature.[49]

Among the Ongee, menstruation is considered a privileged ability of women, given to them by the spirits. An Ongee myth says:

> First the women came from the sea. Then came the men from the forest. Since women came to the land of Ongee before men they were closer and more friendly with the spirits ... Women on reaching a certain age would release blood at each full moon. This capacity to release blood at each full moon was [an ability] given through the spirits ... The capacity of women to bleed was a means for the women to maintain their body always in a state of appropriate distribution of heaviness and lightness within the body.[50]

For the Ongee one's 'heaviness' or 'lightness' refers not to physical weight but to the amount of one's personal odour and vitality. According to this system, menstruation is a natural means of regularly releasing excess odour in order to keep from becoming surcharged with scent.

Ongee men not only lack the ability to regulate their odour through menstruation, they also do not have the close relationship to the spirits that women are reputed to have. On reaching adulthood, therefore, men must compensate for their disabilities by ritually inducing a loss of odour and entering into communication with the spirit world. During this rite of male initiation, the Ongee, contrary to custom, try and attract the spirits by smell. Women and children swing on swings in order to better spread their odour through the air, men go hunting in the forest without any of their usual odour suppressants, and a basket of rotting pig's meat is hung from a tree.[51]

The initiate himself must be made ready to travel with the spirits by having his olfactory weight lessened. In order to accomplish this, women inhale odours from the initiate and induce him to perspire by applying heat packs. The women also massage the initiate, attempting to push his olfactory weight into the lower half of his body and to thereby enable him to float like a spirit.[52]

For two days the initiate is believed to visit and learn from the spirits while his body remains lifeless on the ground. After this period the initiate is cooled down, to solidify his odour, and massaged, this time to redistribute the odour within his body so that he will once again have a stable olfactory weight. This sojourn with the spirits is said to make an Ongee man *gayekwabe* – able to bind or release his odour as the need arises. With his new knowledge and skills the initiate becomes a full-fledged member of the community.[53]

Another rite of passage which often makes use of odour is that of marriage. In the United Arab Emirates, for example, as much money and care is spent on perfuming the bride as on clothing her. The bride's body is rubbed with a blend of rose, musk, ambergris and saffron oils. Particular care is taken with her hair:

> Before the [bride's] hair is braided, a special perfume mixture is combed into the hair. In a pot, musk, saffron and [aromatic leaves] are kneaded in water to form a soft dough which is then incensed with aloewood. For best results, the combination is left a day to blend properly. When the time is appropriate it is placed on the hair with half a bottle of aloewood oil.[54]

The bride's clothes are generously perfumed as well. One method involves soaking the bridal garment for three days in rose-water, saffron, black and white musk and civet, and then placing it on an incense rack and fumigating it with musk and ambergris.[55]

The groom, for his part, incenses his body by stepping several

times over a censer of burning aloewood. He also anoints his
neck, ears, hands and beard with rose and aloewood oils.[56] Such
intensive olfactory preparations are thought to make the bride
and the groom pleasing to each other, to the participants of their
wedding and to God. At the same time they cleanse the bride
and groom of the evil influences of foul odours and enable them
to enter into their new state purified and uplifted.

In Northern Sudan, weddings also involve a heavy use of scents.
A special bridal incense is prepared, of which the major ingredients are sandalwood and *dufra*, said to be the tips of crocodile
claws. 'The [ingredients] for bridal incense are tossed with sugar
and cooked in a pan, then sprinkled with "cold" scent and
smoked perfumes; their fragrance is released by smouldering the
amalgam in a brazier.' This incense is used to fumigate the bride
and her surroundings. Wedding scent, in turn, is composed of oils
of clove, sandalwood and the kernels of a variety of cherry. To
these are added granulated musk, smoked sandalwood powder
and colognes. Finally, tobacco smoke is blown into the bottle
containing this mixture and the bottle is shaken to disperse the
smoke throughout. This scent is used generously by women
attending a wedding ceremony.[57]

Such bridal perfumes would be as familiar a part of a Sudanese
wedding as Wagner's 'Bridal Chorus' is in the West. The use of
scents at a Sudanese wedding, however, is not only a matter
of aesthetics or custom, but of complex ritual symbolism as well.
Commercially prepared colognes and perfumes, for example, are
associated with coldness and with masculinity, while incense is
related to heat and femininity. Blending 'cold' perfumes with
'hot' incense and tobacco smoke in wedding scents creates a unity
of male and female which symbolizes the marital union and
conveys the notion of fertility. At the same time, wedding scents
are an important means of keeping evil jinn away from the ceremony (although, as we have seen above, other classes of jinn
might be attracted by the fragrance). The couple and the participants are thereby protected from supernatural harm during this
vulnerable period of transition by the power of scent.[58]

We have looked here at instances of olfactory symbolism surrounding birth, adulthood and marriage in a variety of cultures.
Other stages of transition, such as becoming a religious specialist
or, as we shall see below, death, commonly make use of olfactory
symbolism as well.

But first, why do odours tend to be emphasized during such rites of passage? The reason would seem to be that there is a widely perceived or intuited intrinsic connection between olfaction and transition. To begin with, it is in the nature of odours to alter and shift, making them an apt symbol for a person undergoing transition. Consider the situation of the initiate at a male puberty rite. The initiate is no longer a boy but not yet a man. He is 'betwixt and between' the conventional categories of social perception. In a similar way, smells are difficult to classify, and even more difficult to contain. Their 'out of placeness' thus corresponds to the ambiguous status of the subject of the rite of passage.[59]

The 'in between' condition of the subjects of a rite of passage makes them particularly vulnerable to the effects of the odours they come into contact with at this time. The wrong odour – for example, a smell of menstruation during a male puberty rite – can prevent the ritual participants from making the necessary transition. The right odours, however, can help guide them across from one stage to the next. In the case of a funeral rite, for instance, the spirit of the deceased might be thought to rise up from earth to heaven with the smoke of burning incense.

Another dimension of the connection between olfaction and transition has to do with the power of smell to bring people together. It is important for the celebrants at a rite of passage to develop a sense of communion. Odour, such as that of incense, helps create the desired fellow-feeling by virtue of the way all concerned become conscious of partaking of and being enveloped in the same smell. This unifying power of smell is especially important when the rite is one of social incorporation – for example, incorporating a newborn into society or uniting a man and woman in marriage. Disparate people, with their different odours, come together and are blended into a social and olfactory whole, though as we have seen, the particular scents used to accomplish this will depend on the code of olfactory meaning of the culture in which the rite of passage takes place.

DO YAMS HAVE NOSES: THE ODOURS OF HUNTING AND GATHERING

While rituals employing odour are often performed for special, out of the ordinary, purposes and occasions, such as rites of passage or communicating with the spirits, they can also deal with everyday concerns, such as procuring food. In traditional cultures, olfactory sympathies and antipathies are often attributed to food resources, whether plant or animal. The Wamira of Papua New Guinea, for example, believe that taros – the tuber which they cultivate – dislike certain odours. These odour aversions convey Wamira notions of the proper relationship between humans and their crops. An ambivalence towards the introduction of new technology into traditional Wamira gardening practice, for instance, is expressed in terms of an olfactory antipathy on the part of the taro:

> Before people used to turn the soil with wooden digging sticks and it was good. There was lots of taro. Now people also use metal garden forks, and some of them have rust. The taro plants disappear because they smell the rust on the forks.[60]

Taros also dislike the odour of any 'oily' substances, such as meat, coconuts and soap, or 'salty' substances, such as fish, seaweed and menstrual blood. The former are associated with male fertility and the latter with female fertility. To bring these odours into the taro fields would be to confuse human fertility and plant fertility – or the process of reproduction (of humans) with the process of production (of taro) – and thereby stunt the growth of the taro.[61]

In the nearby Trobriand Islands, yams, which form a major part of the local diet, are thought to be hypersensitive to odours. The Trobrianders believe that yams are repelled by the smells of cooking, probably because they are deemed reluctant to be cooked themselves. Cooking is therefore prohibited in the area around the storehouse where yams are stockpiled, and small stores of yams, kept close to home for ready use, are protected from offensive cooking odours by being covered with plaited coconut leaves.[62]

Seasonal fruits form an important part of the diet of the Batek Negrito of the Malay Peninsula. Such fruits are so abundant that the Batek can virtually live on fruit alone for two months of the year. To ensure that there will be a plentiful harvest of fruit the Batek meet during the blossoming period to sing fruit songs, a different song for each kind of fruit. These songs emphasize how delicious and attractive the fruit in question is, and encourage the spirits to send down plenty of fruit blossoms. If there is a specialized fruit shaman present, the singing session will climax in his travelling to the land of the spirits while in a trance in order to bring back some extra flowers. When the first wild fruits of the year are harvested, incense is burnt under the fruit tree, offering up fragrance to the spirits in return for the aromatic fruit blossoms they have sent down to the earth. To fail to make this offering is thought to put one at risk of falling from a tree while gathering fruit.[63]

The Batek Negrito supplement their diet with the meat of wild animals such as monkeys, civets, squirrels and bamboo rats. When hunting they are careful not to use any animal skins or feathers as it is thought that the odour of these materials would frighten away game. An exception to this rule is the use of gibbon bones by Batek hunters. In this case curious gibbons are thought to be attracted by the odour of the bones of their fellows, bringing them within range of the hunters' blowpipes. Sometimes animals, particularly tigers, are actually spirits in animal form according to the Batek. In such cases the poisoned darts of the hunters will not harm them. However, it is said that if a person blows incense on the spirit-animal it changes to a human form and imparts knowledge about magical practices. For this reason many Batek Negrito carry incense in their pouch of personal effects so as to be ready in the event of such an encounter.[64]

Among the Desana of Colombia, hunting is the focus of extensive ritual activity, much of it involving odour. Only men are allowed to hunt and they must begin their ritual preparation for hunting in childhood. Before reaching puberty a boy drinks infusions of magical herbs in increasing doses so that his body will become saturated with their aroma. When setting out for a hunt, a man rubs his body and weapons with select aromatic herbs as well. These herbal scents are thought to make the hunter appear attractive and friendly to his game, thereby guiling them into approaching without fear. At the same time, the aroma is thought to act as a general fertility drug, increasing the reproductive potential of the animals who smell it.[65]

The herbal preparations used by hunters are similar to those used by men to attract women. The Desana, in fact, explicitly state that hunting is like courtship. The Desana term for hunting, *vaí-mëra gametarári*, means 'to make love to the animals'. Just as perfume is believed to attract the desired woman to the desirous man, therefore, it is also believed to attract the hunted to the hunter. In either case the ultimate goal is the continuation of the Desana cycle of life.[66]

Apart from rubbing his body with herbs, the Desana hunter takes a number of other olfactory measures to ensure a successful hunt. If he intends to hunt large, strong-smelling 'black' animals, such as tapirs or peccaries, he eats smoked meat and smoked chilli pepper, so that his body will emit the musky odours these animals are thought to find attractive. If he is hunting the preferred mild-smelling 'brown' game – paca, cavi, armadillo – however, he must consume only boiled food and, hours before the hunt, absorb an infusion of chilli pepper through his nose and take emetics in order to cleanse his body of 'residues'. This is supposed to purify the hunter of any foul body odours which would repel the sensitive 'brown' game. Before leaving for the forest, the hunter smokes tobacco, creating an imaginary fence of smoke around his hunting grounds in order to keep unwanted animals, such as poisonous snakes, away. Some of this smoke, which is believed to be very potent, he blows over his weapons to make them more penetrating. [67]

The Desana shaman, in turn, assists the hunting expedition with his supernatural powers. The shaman tries to ensure the desired animals will gather at a certain place by magically placing fresh grass there and sweetening the water with pineapple flavour while in a trance. Traps set for small animals are described by the shaman as delightful gardens full of inviting food. The hunters themselves are depicted to the animals as attractive, pure, perfumed and harmless beings who wish only to pet them. When a hunter has killed an animal, such as a deer or a tapir, its tongue is cut out immediately and buried, and tobacco smoke is blown over the spot. This is thought to prevent the animal from relating how it has been tricked and seeking vengeance.[68]

Returning to the Andaman Islands, the Ongee believe that, just as the spirits hunt *them* by smell, so they hunt animals by smell. The Ongee word for 'to hunt' is *gitekwatebe*, which means to kill by releasing a flow of odour. The Ongee say:

> We have smell so do the animals. We reach the animal and kill it by releasing all of its smell. The winds take the smell away and never does the smell come back – it is death of the animal – success of the hunter – if the hunter lets the wind

take away all his smell he stops moving – it is death – spirits take the hunter away.[69]

The practices of olfactory management which prevent the Ongee from being sniffed out by the spirits who hunt them also serve to keep their presence unknown to the animals who are hunted by them. One odour that would immediately alert the animals to the hunter's intention is the smell of meat emanating from the hunter's body. After eating meat, therefore, the Ongee paint themselves with white clay, which is held to have the virtue of suppressing body odour.[70]

The odour of the bones of hunted animals, however, is said by the Ongee to trick living animals into thinking that their missing fellows are still alive. The animals will then remain in the area without fear. In order to perpetuate this illusion, the Ongee paint the skulls of the animals they have killed with red clay and store them over the cooking area, techniques that are thought to augment the dispersal of odour from the skulls.[71]

This notion of hunting by smell is essential to Ongee identity. The novice who returns from visiting the spirits is re-initiated into the Ongee hunting ethos by shooting a miniature arrow through the nostril of a pig skull. This act symbolizes the killing of wild pigs through the release of their odour-life. The novice is told by his father:

Release the smell of pigs ... again and again – and quickly and with care! Then it will be the turtle whose smell you will release – you then will hide yourself for the honey and then you will hunt the snake, dove and nautilus. One has to kill them all, you have done it – don't forget it because you have to help others to do the same ... Do not be light and afraid. You have to be a hunter, hunting the smells.[72]

These different olfactory beliefs and practices make it clear that odour not only determines the edibility and aesthetics of food in various cultures, but also makes possible the very procurement of food. For the Wamira, the Trobrianders, the Batek Negrito, the Desana and the Ongee, acquiring sufficient food depends on carefully and properly manipulating the olfactory environment. The Wamira and the Trobrianders must guard against introducing any odours detrimental to their crops into the environment of their fields and storehouses. The Batek Negrito must ensure that a sufficient number of aromatic fruit blossoms descend from the spirit world to earth and then, to have a safe harvest, thank the spirits for their generosity by sending up the fragrance of incense. The Desana try and attract the animals they hunt for food by guiling them with perfumes. Amid the multitudinous odours of the forest, the desired prey is believed to pick out the seductive scents exhaled by the hunter and inexorably make its way to him. For the Ongee, hunting for food is a game of olfactory hide-and-seek. A successful hunt involves restricting the release of the hunter's odour and locating and releasing all the odour of the prey.

These practices are based on the belief that animals and plants are part of the same olfactory network which interrelates humans and spirits. This is stated most clearly in the case of the Ongee, where the way humans take away the life-smell of animals parallels the way spirits take away the life-smell of humans. Furthermore, many of the foods eaten by the Ongee are thought to contain spirits who happen to be feeding on those foods at the same time. This poses a moral dilemma for the Ongee, who say: 'We cannot do wrong to our ancestors who are dead by eating the food which they are eating and are inside. However we cannot do wrong to our living relatives by not eating anything.'[73] As it is impossible to keep food resources and spirits completely separate, the Ongee must sometimes consume spirits with their dinner. This act, however, is not all negative as the consumed spirits are thought to turn into human babies.[74]

Animals and plants are not only interrelated with humans and spirits in many cultures; they are often perceived *as* disguised humans or spirits. The Wamira and the Trobrianders, for instance, attribute human qualities to the tubers they grow. 'Taro are people,' say the Wamira.[75] The Batek Negrito hold that certain animals are actually spirits in animal form. A spirit in the form of a monkey, for instance, is said to protect the celestial store of fruit blossoms from raids by Batek shamans by pelting the offenders with prickly jungle fruits.[76] The Desana believe that their shamans must offer up human souls in payment for the animals they kill. Parents who have lost a child, therefore, will blame the shaman, saying, 'He pays for our food with the lives of our children.'[77] These ransomed souls, rather than returning to the realm of the spirits, become animals, who, again, are hunted by the Desana.

The activities of hunting, gardening and gathering thus resound in many different domains at once, so that yams have noses, forest animals are perfume-loving humans, and the grounds where one hunts animals by smell are also the grounds where the spirits hunt one by smell. The olfactory and other rituals undertaken during these activities are directed towards successfully interacting with and manipulating these shifting domains of existence in order to transform one part of them into food for oneself and one's community.

HERBAL TEA AND MEDICINAL SNUFF: AROMATHERAPIES

One of the most common ritual uses of odour across cultures is to combat illness. This use of odour is typically based on a simple proposition: that harmful odours produce disease and that healthful odours will drive out these harmful odours and thereby cure the patient. A well-elaborated example of this common theory of aromatherapy occurs among the Warao people of Eastern Venezuela. The Warao believe that disease is caused by putrid odours, such as are found in swamps. These fetid odours are said to seek out persons who are weak, pervading their bodies and making them ill. Warao shamans therefore say:

> Fetid air is not of the patient. It comes to him when weak and stays about him. Similarly, a fish while alive, has no bad odor. But when pulled out of the water its life essence fails and fetid air seeks it out.[78]

Warao herbalists, who are traditionally women, combat these fetid odours with fragrant herbal remedies. While fetid odours are thought to have their source in the foul underworld located in the west, fragrance is believed to originate in the refreshing land of the God of Life in the east. The fragrance employed for healing purposes is carefully chosen and blended from a wide selection of aromatic plants. Herbalists are thus also perfumers, creating healing scents. The aroma of the herbal perfume is thought to enter the patient's body and oust the noxious odours. When the process is complete, the herbal aromas depart and the patient returns to a normal, inodorate, state of health.[79]

The Batek Negrito similarly hold that odours can both cause and cure disease. One form of cure involves rubbing the leaves of plants with medicinal properties between one's hands and then smearing the odorous sap on the skin of the patient. Another consists of blowing incense smoke on the afflicted parts of the patient's body. The fragrant incense is supposed to attract the harmful odours, so that they follow the smoke away from the patient. Yet another healing technique is to drink human blood (usually obtained from the leg of a volunteer). In this case the bad smell of the blood is said to drive out the disease.[80]

In the Andes, fragrant herbal teas, which work by both being inhaled and being ingested, are used along with incense to heal and prevent illness. *Muña muña* (*micromria eugenoides*), an Andean shrub with a strong musky smell, is made into tea to alleviate stomach aches. *Muña muña* is also thought to be a potent aphrodisiac – its name comes from the Quechua word *muñay*, to love. Tea made from *vira vira* (*gnaphalium vira vira*), a spicy-smelling plant, is said to help chest colds and coughs, and to improve blood circulation. Eucalyptus vapours and fumigations are also used to treat patients with colds, while fumigations with thorn apple are thought to help in cases of asthma. One of the most important medicinal plants in the region is rue. Rue tea is recommended for nervousness and stomach aches. Rue vinegar, rubbed on the forehead, is said to help against dizziness. Rue is also a basic ingredient in the creation of various types of healing incenses. So great is the belief in the protective powers of rue that many inhabitants of this region will keep a rue plant outside their front door to ward off any ills.[81]

Variations on such 'aromatherapeutic' beliefs and practices can be found around the world. There also exist some highly unique indigenous theories of the role of smell in health and illness. The Ongee, for instance, perceive health as being a balanced state of body odour, and illness as being a decrease or excess of body odour. The decrease is held to be caused primarily by heat and the excess by cold. Heat is said to melt the solid smell contained in the skeleton so that it is transformed into liquid and emitted through the skin. This results in a loss of weight. An overheated Ongee will thus say, 'I am so hot that I will burn down like ash and become "light".'[82] Coldness, on the other hand, is thought to condense odour within the body, producing an increase in weight. In this situation the Ongee say, 'I am shivering, on becoming cold I will fall down like a heavy thing in the sea!'[83] The danger in the former condition is that the spirits will track down the overly odorous patient and take him or her away. The danger in the latter is that the patient loses her or his ability to communicate through the regulated emission of body odours, and thus can no longer remain in contact with helpful ancestral spirits.

The Ongee treat heat-producing illnesses by employing methods thought to stop the release of odour. These include cooling the patient's body and painting it with white clay, considered to restrict the emission of odour. Cold-producing illnesses are treated in the opposite fashion; heat packs and 'hot' red clay are applied to the patient to stimulate odour emission. The balance between smell contained within the body and smell released by the body is thus restored.[84]

Among the Temiar of Malaysia, an olfactory illness is produced when one person passes without warning behind another person's back. Such an action disturbs the odour-soul believed to reside in a person's back, and this odour-soul then proceeds to 'eat' the passer-by, causing illness. Illness can also occur when a person comes into contact with odours which have been left behind by visitors or departing residents. The Temiar say that illnesses caused by odour are characterized by a hard, swollen stomach, gas, vomiting and diarrhoea. In order to cure this malady one takes a cloth containing some of the same odour as that which is causing the illness, rubs it on the patient's stomach to draw out the intrusive odour and then burns the cloth so as to startle the odour into leaving with the smoke. The Temiar are understandably concerned to ensure that treatment will be available should anyone fall sick from an odour left behind by visitors. For this reason, visitors are typically asked to deposit some of their odour on a cloth by rubbing it over their bodies before departing.[85]

The odours employed to heal can be symbolic as well as actual. Among the Shipibo-Conibo of the Peruvian Amazon, patients are treated with fragrant herbs and tobacco smoke, but also with aromatic songs. These songs are imagined to 'fizz' with fragrant gas like yucca beer and also to contain the images of certain traditional geometric designs. When they are sung by the Shipibo-Conibo shaman, they float down onto the patient, reordering the disordered body by the power of their aroma and according to the geometric patterns they embody.[86]

Odours may also play a role in the treatment of psychological ailments. Fragrant herbal teas, baths and incense are used in various Latin American regions to cure everything from depression to bad luck in love or business. An Andean, for example, who is suffering the after-effects of a severe fright may be treated with an incense of rue, copal, earth, sugar and rosemary. Among the Zaramo of Tanzania, bad luck is thought to be the result of being contaminated with an evil odour, known as *nuksi*. This smell is said to be the cause of such things as not finding work or going to prison. Persons suffering from *nuksi* are treated with medicinal plants boiled in water. This treatment takes place on the site of a rubbish dump – the proper place for *nuksi* to reside.[87]

Evil spirits are often held to be the source of harmful odours and illnesses, and aromatherapies are directed towards combating these spirits, as well as the illness itself. Among the Hausa of Northern Nigeria and surrounding regions, witches are thought to enter the soul of an individual through his or her nose, causing mental derangement. Individuals manifesting symptoms of mental illness are therefore commonly treated with hot, peppery medicinal snuff which, by causing violent sneezing, is thought to expel the offending spirit. As a supplement to this treatment, the afflicted person will be placed in a small room, which is then filled with a thick, disagreeable smoke of burning shrew mice, scorpions and herbs. It is hoped that the spirit will be made so miserable by this unpleasant fumigation that it will soon depart.[88]

These different aromatherapies all treat odours as potent forces for health or illness. Therapeutic odours are customarily thought to act on the body by driving out or nullifying the illness. The odours used for this purpose vary widely, from herbal scents through body odours to acrid smoke, according to the culture and the disease being treated. Illnesses themselves are often believed to be caused by foul odours. The assumption in such cases is usually that odours will convey the essence of whatever they emanate from to whomever inhales them. By inhaling a putrid odour, therefore, one invites the process of corruption into one's own body, while inhaling a cleansing odour purifies.

Healing and illness are often considered to have their basis in the supernatural, and this also applies to the odours associated with them. Supernatural odours may cause illness, and such odours, invoked by the healer, may be employed in the battle against illness. Even when it is a matter of actual scents, such as swamp odours or herbal incense, these are often thought to acquire their power from a supernatural source.

Apart from whatever physical or supernatural good they are considered to do, aromatherapies may also have important psychological effects. The Shipibo-Conibo, for example, say that placing a patient in a pleasant, fragrant ambience improves the person's sense of well-being.[89] Likewise, an Andean who has been censed as a treatment for shock, or a Zaramo who has been purified of the lingering smell of bad luck with medicinal herbs, will feel reassured that his or her problem has been dealt with in a tangible, or rather 'smellable', fashion. In the case of the Hausa treatment for a person suffering from mental illness, at the same time as the peppery snuff and pungent smoke are making the invasive spirit's life miserable, they are providing a strong motive for the patient to cease manifesting the symptoms which provoked the treatment. Aromatherapies thus act on three different levels – the physical, the supernatural and the psychological – which overlap and mingle to fulfil their healing functions.

THE DANCE OF OUR ROTTED MEN: FUNEREAL ODOURS

Death is an event of such magnitude in human experience that there is probably no culture on earth that does not ritualize to some degree the deaths of its members. Odour, for various reasons, is often an essential ingredient of mortuary rites. The odour of the cadaver is a potent sign of the end of the life of the body and can consequently serve as the basis for a cultural discourse dealing with life, death and the afterlife. Perfumes, such as of flowers or incense, when used at a funeral, help mask the decaying odour of the body. At the same time perfumes may be thought of as helping the soul of the deceased on its spiritual journey, by rendering it agreeably fragrant to the deities, for example, or by allowing it to climb up with the incense smoke. Funereal odours in general, whatever their source, help mark the rites of death as extraordinary occurrences, olfactory ruptures in the varying sensory patterns of day-to-day life.

Among the Batek Negrito, the bodies of the dead are traditionally placed in trees. This treatment is accorded to the bodies both of humans and of those monkeys and gibbons that have been adopted as babies by the Batek and raised within the Batek community. Among some groups children who die soon after birth are buried in the ground, as it is thought that the body is too tiny to be placed in a tree and that the baby might be afraid of falling down during a storm. In a tree-burial, the corpse is dressed in a colourful sarong, wrapped in a sleeping mat, and laid inside a lean-to built in the branches of a sturdy hardwood tree. Incense resin is then burnt under the tree by the head of the corpse. This is supposed to enable the soul, already located above the earth in the tree, to waft to the afterworld on the incense smoke. The grave site is usually avoided after this ceremony, for it is imagined that the odour of the corpse will attract tigers, who will wait under the tree for pieces of flesh to fall down.[90]

The Batek Negrito characterize the afterworld as a vast, sandy plain, bathed in a cool light and filled with fruit blossoms. The Batek dead, who have fragrant cool dew in place of foul warm blood in their veins, spend their time singing and decorating themselves with flowers. The dead of other peoples and eras, however, are not as fortunate, for they are said to live in a foul-smelling land under the western horizon.[91]

In the United Arab Emirates perfume is a central element of funeral rites. The body is washed with sweet-smelling leaves and scented with sandalwood, camphor and saffron oils. Two censers, in which frankincense and gum arabic are burnt, are placed at either end of the body. The burial shroud is incensed with aloe-wood sticks. The reason given for perfuming the corpse is that the deceased should have a pleasant smell on encountering God. Perfumes would also serve to attract angels to the corpse and dispel evil spirits. None of the family or visitors wear perfume or incensed clothes, however, nor do the customary rounds of incense and perfume rituals follow the drinking of coffee. This is not only due to the ascetic requirements of mourning but also to mark a separation between the living and the dead. Wearing perfume would, in this case, identify one with the corpse and is therefore considered dangerous and an action which could lead to other deaths. Furthermore, the unilateral wearing of perfume by the corpse signifies that the deceased can no longer participate in the olfactory exchange which characterizes the social relations of the living. Once the body is buried in the cemetery, its tomb is rarely, if ever, visited. This is probably due, in part, to a belief that the cemetery, as a site of death and decay, is inhabited by evil jinn.[92]

While grave sites are avoided by the Batek Negrito and the

inhabitants of the UAE, among the Brazilian Bororo, a grave is
not only visited after the funeral, but it is even dug up, to check
the state of decomposition of the deceased. This is done in order
to ascertain how much *jerimaga*, the putrid smell of the life force,
remains. When the odour has almost gone, the body is said to be
'done'. This is the same expression used by Bororo cooks in
relation to meat which has been boiled to rid it of its *jerimaga*.
A corpse which is 'done' is one from which the flesh has rotted
away, leaving only the skeleton, representative of pure form. The
soul is now said to be freed from the bonds of organic life to
become a fragrant wind and to join its ancestors and the 'master
souls' of all species in the spirit world.[93]

The Ongee also attach symbolic importance to the bones of
their dead. For the Ongee, however, bones are condensed smell
and condensed life energy. The bodies of their dead are disinterred, therefore, not to ensure that their vital odour has
departed, but to recover the odour-force contained in their bones.

The funerary rites of the Ongee are intimately bound up with
their conception of the cycle between humans and spirits. The
Ongee believe that the spirits of their dead are reincarnated as
children. Infants are said to be like the odourless, boneless spirits
because their bones are soft and they lack teeth. Ongee children,
in fact, are not considered fully human until their teeth appear.
(Due to their recent experience of being a spirit, however,
children are deemed to have more knowledge of the spirit world
than their elders. The latter will therefore reply when perplexed
by a difficult question: 'The elders cannot answer this question,
ask somebody younger.')[94]

The Ongee associate the onset of death with the loss of teeth,
the loss of condensed odour-force. Elderly Ongee will say, for
example, 'Our teeth are falling, we grow old and die.'[95] The Ongee
bury their dead in the earth. On the first night of the full moon
following the burial, the body is dug up and the lower jawbone
is recovered along with other bones. A lower jawbone with
teeth is called *ibeedange* – dangerous smell body – by the Ongee.
This is because chewing meat, like killing a human or animal, is
thought to release dangerous odours. By removing the jawbone
from a body the Ongee ensure that the deceased's spirit will be
unable to chew, and thus be less aggressive in hunting and more
willing to cooperate with humans.[96]

Bringing the bones back to the Ongee camp signals the end of the period of mourning. The close relatives of the deceased tie dry plants around the bones to cool them and keep their smell in, and paint them with red clay so that the bones will not be too cold and will continue to emit odour. Finally, string is tied to the bones so they can be worn on the body. The bones are kept in a basket by the family and provide a means of maintaining ties with one's ancestral spirits through smell. On occasions of great need, such as when a family member is ill, they are taken out and worn on the body. The odour of the jawbone, mingled with that of the wearer's body, would serve to alert the ancestral spirit of the need for his or her intervention.[97]

The Ongee imagine the spirits of the dead to be odourless. The inhabitants of certain Melanesian islands, such as New Caledonia, however, assign a putrid odour to the spirits of their dead. It is said that none can enter the land of the dead who do not manifest this scent. A living human who wishes to visit there, consequently, must first anoint himself or herself with the decaying remains of a dead animal. In New Caledonia it is thought that the recently dead still smell of life when they enter the underworld. This alien odour disturbs the spirits already there. They throw the newcomer a bit of their food to eat, an action which causes all of his or her offensive odour of life to disappear. Thus, whereas the food of the living is life-giving, the food of the dead confers the state of death.[98]

In New Caledonia the dead are said to spend their time in rhythmic activities, playing ball with an orange, changing their body colour in unison – from white to red to black and other colours – and dancing over arid plains and mountains with trees and rocks, which also provide them with temporary abodes. The living imitate this dance of the dead when they celebrate the end of the mourning period, three or four years after a person has died. The leader convokes the participants: 'Rise all of you, come for the dance of our rotted men, smelling of rancid fat, who live in the holes in rocks and the trunks of trees.'[99] The women and men dance around a pole all night, heavily, rhythmically, as the dead do.[100]

In nearby Northern New Ireland, funeral rites involve a careful transference and dispersal of odour. Life force is said to manifest itself in humans as smell and to increase with age. At death this odour of life slowly leaves the body. The odour of life is thought to be dangerous when not contained within a body, however, so the New Irelanders create a sculpture in order to capture the smell emanating from the corpse. The type of material used – wood, fibre or clay – depends on the amount of smell the deceased is believed to have accumulated. An elderly man, for example, is considered to have the greatest store of life-smell and will therefore, according to custom, be represented by a sculpture made of wood when he dies.[101]

As it takes on the odour emanating from the deceased the sculpture is said to grow alive. When the sculpture is displayed publicly its particular design is carefully memorized by certain individuals. The sculpture itself is then left to 'die' and disperse its now devitalized odour. A wooden sculpture is left to rot and exude its odour through decay. A fibre sculpture is burnt, releasing its odour in the smoke. A clay sculpture is deprived of its acquired odour by being taken apart. The funeral rite is now complete.[102]

The various olfactory practices and beliefs surrounding death presented here reveal the different means by which societies try to make cultural order out of the disorder of death. In the modern West the odour of the corpse is suppressed through techniques of embalming in order to reduce the trauma of death for the survivors. For the inhabitants of the UAE, perfumes make the deceased presentable, both to the mourners and to God. Once the perfumed spirit has departed for heaven, however, the decaying buried remains become a site of danger, to be avoided by the living. The Batek Negrito believe that incense aids the spirit to depart from the body. The spirit then becomes a fragrant superhuman, while the decomposing body attracts dangerous tigers by its odour. The inhabitants of New Caledonia, in contrast, believe that the spirits of the dead smell of their decaying corpses. In this case, the odour of death is also the odour of the gods.[103]

While the odour of the corpse is identified with death in many cultures, it can, interestingly, also be identified with life. The Bororo, who hold that the life force has a putrid smell, can readily conceive of life as departing from the body in olfactory form as the corpse decomposes. Like the Batek Negrito, the Bororo believe in an olfactory separation of body and soul at death: the soul is said to become a fragrant wind after the body has released all of its stench of life through putrefaction.

For the Ongee, the spirits, divested of the odour of life, are inodorate. With regard to their human remains, it is not the odour of decaying flesh that the Ongee emphasize, but rather the odour of bones. These bones, as sources of condensed smell and condensed life energy, are kept and utilized by the Ongee in order to transmit olfactory messages to their ancestral spirits.

Finally, the people of Northern New Ireland channel the odour of life departing from the corpse into a specially designed sculpture. Once the design of the sculpture is committed to memory, the sculpture itself is destroyed, its transformative function fulfilled. In this way, the New Irelanders are able to convert the transient smell of life and of the deceased into a fixed visual image, a sort of olfactory 'photograph', to be preserved indefinitely.[104]

SCENTED DREAMS: THE ROLE OF SMELL IN DREAMS AND VISIONS

In many cultures odours play a ritual role in the production of dreams or visions. Among the Umeda of Papua New Guinea the word for dream (*yinugwi*) is very similar to that for smell (*nugwi*). Perhaps due to this perceived similarity, an Umeda man always sleeps with a sachet of ginger by his side or under his head. It is believed that the scent of ginger will stimulate dreams which will augur well for hunting. Just as the word for dream is similar to that for smell in the Umeda language, so is the word for ginger (*sap*), the pre-eminent magical herb, a synonym for magic. Thus, in the context of the dream, the magical odour of ginger acts upon the imagination of the dreamer to produce a prophetic vision which will alter the world in favour of the dreamer.[105]

Among the Ongee the role of odour in dreams is essential to the continuance of life. The Ongee believe that a person's spirit resides in her or his bones. During the night, while a person sleeps, this spirit goes out and visits the places the person has been during the course of the day in order to collect the smell which has been left behind and bring it back to the body. This scent-gathering is experienced by the individual as the process of dreaming. The continual restoration of odour-life to the body during sleep replenishes a person's depleted vitality and enables her or him to continue taking in and emitting smells. When death occurs the Ongee say that it was because the individual's smell did not return and that she or he therefore had 'nothing to breathe'.[106]

In the hallucinogen-taking cultures of South America, odours are used to control and direct visionary experience and also take on special meaning when they appear during visions. The Amazonian Amahuaca, for example, envelop themselves in a fragrant smoke before ingesting the hallucinogen *ayahuasca*, in order to induce a sense of tranquility. The hallucinations themselves are directed by the shaman who makes use of them to explain the life of the forest in full sensory detail. A Peruvian rubber trader who was captured by the Amahuaca and lived with them for many years describes the hallucinatory experience as follows:

> I could separate the individual notes of the bird song and savor each in its turn. As the notes of the song were repeated, I floated in a sensation that seemed somewhere between smelling an elusive intoxicating fragrance and tasting a delicate ambrosia. A breath of cool air drifting in from the forest created an ecstacy of sensations as it cooled my exposed skin. Sensations of a pleasant aroma again seemed involved.[107]

What is described here is the sensation of synaesthesia, or mingling of perceptions, that taking ayahuasca produces. This hallucinogenic synaesthesia has a great deal to do with the ways in which the senses are interrelated by *ayahuasca*-taking cultures such as the Amahuaca or the Desana. The Desana, indeed, place the hallucinogenic experience at the core of their culture, saying that it reveals ultimate truths about cosmic reality.

Only initiated men are allowed to take *ayahuasca* in Desana culture. The ritual begins after sunset and takes place in the communal house. The men, who have prepared for the ritual by undergoing a period of sensory deprivation, drink a potion of *ayahuasca* out of gourd cups. Recitations by the Desana shaman alternate with music and dancing and the smoking of tobacco. All sensory cues – the shake of a seed rattle, the red glow of the fire, the smell of smoke – are said to influence the form of the hallucinatory experience.[108]

The Desana shaman serves as a guide through the visionary world. At first the drug taker is said to feel as though he is flying up to the Milky Way. The Milky Way is conceptualized by the Desana as a yellow skein of palm fibres, a vast seminal flow, or a foaming river full of putrid residues. It contains both fertilizing and contaminating essences. While flying through the Milky Way the drug taker perceives luminous yellow 'stars and flowers'. When he has passed beyond the Milky Way he sees the super-natural beings of the Desana cosmos – the Sun-Father, the Master of Animals, Thunder-Person – and the myths of creation are enacted. Game animals crowd the scene, accusing the Desana of killing them and clamouring for justice. The moving forms and colours now swirl and transform into 'wide open scenes of placid clouds bathed in soft green light'.[109] The drug taker, enveloped in light and floating on waves of music, is lost in contemplation. This is the Desana afterworld, the place where souls go after death. It is known as a 'land of milk and coca', and the green light which illuminates it is said to be the colour of young coca leaves – requiters of hunger and thirst. It is also the womb, the place where all one's needs are satisfied.[110]

Desana shamans say that in directing the multisensory halluci-natory experiences of their followers they have four basic goals in mind: 'to make one see, and act accordingly'; 'to make one hear, and act accordingly'; 'to make one smell, and act accord-ingly'; and 'to make one dream, and act accordingly'.[111] The experiences gained by taking hallucinogens therefore must trans-late not only into knowledge, but also into proper social behaviour. At the same time, these meaning-rich hallucinatory images transfer their significance onto everyday sensory images – a bird call, the scent of a flower, the pattern of a honeycomb – reminding the Desana at all times of the moral values and governing principles of their cosmos.

The Desana also find meaning in the sensory imagery of dreams. A Desana shaman, for example, will be forewarned of the arrival of visitors to the settlement by dreaming of the event in advance. He is able to determine which tribe the visitors belong to by the scent of the particular aromatic herbs their dream counterparts carry in their belts.[112] Even in dreams, therefore, odour serves to identify and classify among the Desana.

In all of these examples odours act as triggers and/or subjects of dreams or hallucinations. For the Umeda, odours stimulate dreams which are themselves like odours. For the Ongee, dreams are times of odour-gathering, of restoring the depleted olfactory strength of the sleeper. For the Amahuaca and the Desana, hal-lucinations are guided in part by odour and reveal odours to be part of a complex web of sensory and social meaning.

Odours and dreams can be perceived as alike in many ways.

Both are tangible and transitory. Both also can provide knowledge beyond that of the visible world, conveying essences hidden to the eye. In the modern West odours are commonly thought to play very little, if any, role in dream-life. The Umeda, Ongee, Amahuaca, Desana and many other peoples know differently.

Part III

Odour, power and society

Chapter 5

Odour and power
The politics of smell

Due to its marginal and repressed status in contemporary Western culture, smell is hardly ever considered as a political vehicle or a medium for the expression of class allegiances and struggles. None the less, olfaction does indeed enter into the construction of relations of power in our society, on both popular and institutional levels. In keeping with the modern regime of olfactory silence, the centre (the power elite) governs from a position of olfactory neutrality. Formerly power was personal, and therefore imbued with the smell of those who wielded it; now it has become impersonal and abstract, and therefore odourless.

While groups in the centre – politicians, businessmen – are characterized by a symbolic lack of scent, those on the periphery are classified as odorous. Women, for example, are either 'fragrant', or 'foul'. Ethnic groups exude 'foreign', 'undesirable' odours. The working classes, in turn, 'reek' of poverty and coarseness. The olfactory challenge for those in power is how to preserve their inodorateness from the onslaught of odours emanating from these peripheral groups which always seem to be pressing in towards the centre. The challenge for the periphery is two-tiered: on one level marginalized groups internalize their olfactory classification and attempt to gain respectability by dispelling or masking their presumed ill odour; on another, such groups seek to assert their own olfactory norms, evaluating their olfactory identity as positive and denouncing the false olfactory identity foisted on them by those in power.

In this chapter we shall examine how such olfactory codes create and inform power relations between social, classes, ethnic groups, and women and men in the contemporary West. We will also be discussing the regulation of odours in public space and some of the controversies surrounding our 'right to clean air' and our 'freedom of smell'. The final section, on 'The Fragrant Utopia', explores how the political role of odour in three literary utopias of the twentieth century both reflects and critiques the structures and trends presented above.

THE ESSENCE OF GENDER

It is a commonplace in the modern West that women are the perfumed sex. Indeed, the main purpose of perfume, according to a book on perfumery published in 1940, is 'to make women more attractive and alluring'.[1] Thus, by the mid-twentieth century, all the premodern uses of fragrance have come down to one, that of enhancing women's desirability for men.

When we examine the myth of the perfumed sex more closely, however, we find that it breaks down into several stereotypical olfactory and feminine categories. Certain women are not considered fragrant at all in Western tradition. These include prostitutes, viragos, and virtually any woman who defies the established, male-domimated social order. Such women are bad odours on the olfactory scale of feminine value. (Thus the Spanish word for whore, *puta*, along with the French *putain*, are derived from the Latin for putrid.) Maidens, innocent and docile, on the other hand, are naturally fragrant and should smell of nothing stronger than the flowers with which they are associated. Wives and mothers are surrounded by smells of cooking, with a dash of some respectable perfume, nothing too heady, thrown in on special occasions. Seductresses are *femmes fatales*, sirens who lure men to their deaths. They symbolize dangerous attractiveness and their scents are heavy and spicy.[2]

The modern Western man, however, if a *real* man, disdains the use of scent, along with all other cosmetic artifice. On this our 1940s author, A. H. Verrill, decisively and, one senses, proudly, states:

> Though his women use more perfumes than any other race on earth, and although the inhabitants of the United States spend more money on perfumes and cosmetics than on education, the use of perfume in any form is abhorrent to the American male. In vain have perfumers tried to introduce 'manly' scents such as leather, scotch, hay, clover, and salt marsh, etc. Not

one has succeeded. And the men of American blood remain
firm in their determination not to use perfumes.[3]

Men are the pursuers of women, they are the choosers, not the
chosen, and as such there is no particular need for them to
enhance their attractiveness. It is women's duty, however, to exert
themselves to appear attractive to men, attractive enough to be
pursued and chosen. They are prey who must leave scent trails
for their hunters. In undertaking this duty, women run several
risks. First of all, they must use devices, such as perfumes and
high-heeled shoes, considered somewhat ridiculous, or at least
impractical and frivolous, by men. They thereby, at the same
time as they attempt to enhance their attractiveness, surround
themselves with an aura of foolishness. No *sensible* person (i.e.
man) would wear such nonsense.

On the one hand, men regard these feminine embellishments
as being undertaken primarily on their behalf. On the other,
however, they see them as somehow essential to feminine nature,
an innate female perversity. Verrill writes:

> They would use perfumes even if by doing so they lost their
> husbands, and they would willingly suffer the tortures of dun-
> geon and pillory rather than abjure the tortures of high-heeled
> shoes and the operations of beauty specialists.[4]

There is no need, therefore, for men to sympathize with the
troubles women take to conform to the current cultural ideal of
feminine beauty, for 'whoever heard of a woman complaining
of pain or discomfort occasioned by the process of adding to her
personal charms?'[5]

Furthermore, once they have enhanced their attractiveness by
all the means at their disposal, women lay themselves open to
the charge of employing artifice to deceive men. Indeed, an English act of 1770 went so far as to enable criminal charges to
be laid against any woman who offended in this way. The act
provided:

> That all women, of whatever age, rank, profession or degree,
> whether virgins, maids or widows, that shall from and after
> this act impose upon, seduce and betray into matrimony any
> of His Majesty's subjects by the use of scents, paints, cosmetic
> washes ... shall incur the penalty of the law now in force
> against witchcraft.[6]

Here is the myth of woman as dangerous seductress, using perfume to lure men as witches use black magic.

If using scent makes women witchlike, however, not using scent also renders them witchlike – for then they run the risk of being perceived as malodorous. While men are allowed to smell sweaty and unpleasant without losing any of their masculine identity, women who don't smell sweet are traitors to the ideal of femininity and objects of disgust. This is all the more true in that, behind the myth of the perfumed sex, women are suspected of being naturally foul, reeking of unpleasant body fluids, such as menstrual blood.

No one has expressed this traditional belief more strongly than Jonathan Swift. In his poem 'The Lady's Dressing Room', a man discovers all the foul odours – of dirty clothes, chamber-pots, and so on – which underlie the elegant front his beloved presents in public. After this olfactory revelation:

> His foul Imagination links
> Each Dame he sees with all her Stinks:
> And, if unsav'ry Odours fly,
> Conceives a Lady standing by.[7]

The poem ends by instructing men to dwell not on the 'reality' of women's foulness but on the 'illusion' of their beauty. Women, in turn, are advised by Swift in another, similar poem to conceal the foul aspects of their physical nature as much as possible in order to avoid disillusioning men.[8]

This complex of olfactory representations concerning women continues to hold sway in the late twentieth century, in spite of protests made against it.[9] It is, indeed, an important part of what makes the marketing of deodorant as well as perfume products for women such big business.

While the customary use of perfumes by women is clearly bound up with sexist beliefs and practices, this custom has none the less allowed women to gratify and stimulate their sense of smell in ways that twentieth-century men have been socially unable to. As we shall see in the next chapter, however, the aggressive promotion of perfumes for men in recent years may put an end to this sex difference by making a predilection for scent one of the identifying characteristics of the 'red-blooded American male' (*pace* A. H. Verrill).

CLASS AND ETHNIC ODOURS

Different odours are often ascribed to different social classes and ethnic groups in the West. Variations in group odours may be caused by such things as differences of diet, hygiene and perfume practices. With reference to diet, for example, Richard Doty writes: 'Distinctive odours attributed to Indians are probably the result of their eating sweet relishes and spices. Lean mutton has been suggested as the cause of the odour of Arabs.'[10] The belief that food affects body odour has a long history in the West. In the eighteenth century Benjamin Franklin had the following to say on the olfactory effects of one's diet:

> He that dines on stale fish, especially with much addition of onions, shall be able to afford a stink that no company can tolerate; while he that has lived for some time on vegetables alone, shall have that breath so pure as to be insensible of the most delicate noses.[11]

Often, however, a given ethnic or class odour is considered not just to be due to the consumption of particular foods or to perfume practices, but to be somehow intrinsic to the group, a characteristic trait as inalterable as skin colour.[12] Such 'ethnic' or 'racial odours' are commonly portrayed as both distinctive and disagreeable by those people who make an issue of them. The same people normally invoke such odours to justify avoidance behaviour. By way of example, an unemployed youth from Birmingham, England, had this to say about Pakistani immigrants to the UK: 'I just don't like Pakis. They stink. Pakis really reek. You can tell one in the street a mile away.'[13]

We see in the Birmingham youth's remarks a typical example of racist olfactory discourse. It is rarely one's own group which smells – for just as one tends not to notice one's own body odour, one tends to regard one's own social group as inodorate – it is always 'other people'. As the sequence of the youth's statements – 'I don't like Pakis. They stink' – further reveals, a feeling of dislike towards a given class of people usually precedes and informs the perception of them as foul-smelling. Rather than a cause of ethnic antipathy, therefore, olfactory aversions are generally an expression of it.

The odour of the immigrant has become the subject of high-level political debate in European countries in recent years. In

France, former prime minister Jacques Chirac, seeking to win a share of the so-called anti-immigrant vote, professed sympathy for the French worker, whom he characterized as forced to put up with the noise and smell of the immigrant family living off welfare next door. Groups traditionally associated with the political left have denounced this vote-grabbing ploy as inciting racial hatred.[14]

These debates reflect the changing ethnic composition of European society, for traditionally in Europe it was not so much ethnic groups but the working class which 'stank' – from the perspective of the middle and upper classes. In the 1930s, George Orwell went so far as to suggest that 'the real secret of class distinctions in the West' can be summed up in 'four frightful words ... *The lower classes smell.*'[15] The stench of the working-class body constituted an 'impassable barrier', in Orwell's experience, to close association between the classes:

> For no feeling of like or dislike is quite so fundamental as a *physical* feeling. Race-hatred, religious hatred, differences of education, of temperament, of intellect, even differences of moral code, can be got over; but physical repulsion cannot.[16]

This perceived malodour of the working class was due in part to the fact that, as manual labourers, workers naturally perspired when they worked (unlike the bourgeoisie, whose work usually involved little physical exertion) and to the lack of bathing facilities in working-class homes. Somerset Maugham drew out some of the social and political consequences of such unequal access to the 'means of sanitation', as it were:

> The matutinal tub divides the classes more effectually than birth, wealth, or education ... I venture to think that the cesspool is more necessary to democracy than parliamentary institutions. The invention of the 'sanitary convenience' has destroyed the sense of equality in men. It is responsible for class hatred much more than the monopoly of capital in the hands of the few.[17]

Orwell likewise believed that practices of personal hygiene served to divide the classes to a far greater extent than was perhaps commonly thought or admitted. He recalled how a middle-class consciousness was instilled in him as a boy through his being 'taught almost simultaneously to wash his neck, to be ready to die for his country, and to despise the "lower classes" '.[18]

At the same time, Orwell recognized that the perception of class differences went beyond hygienic practices, for as he observed (again drawing on personal experience):

> even 'lower class' people whom you knew to be quite clean – servants, for instance – were faintly unappetising. The smell of their sweat, the very texture of their skins, were mysteriously different from yours.[19]

The malodour of the working class was not so much an actual smell due to poor hygiene, Orwell is suggesting, as a feeling on the part of the bourgeoisie that workers and servants, because of their low, 'foreign' status, were morally and physically repulsive.

We find a similar cluster of representations and attitudes in the United States concerning blacks. The social psychologist John Dollard noted in *Caste and Class in a Southern Town* that whites often claimed that blacks had a 'disagreeable', 'acrid' smell which made close association between the races impossible. This odour was thought to be not simply due to living conditions but intrinsic to blacks, whether labourers or middle class, and noticeable among 'even the cleanest of them'.[20]

Dollard confessed that he was unable to detect any categorical difference between the body odour of black labourers and white labourers, and that the alleged body odours of middle-class blacks completely escaped his nostrils. This absence of any empirical support for the belief that blacks have a distinct and disagreeable smell led him to surmise that 'the odour point', as he dubbed it,

> is greatly overworked and I consider it even possible that the widespread existence of the belief itself may induce a hyperfastidious sensitivity toward Negro body odor which is not displayed toward the body odors of white people.[21]

It is noteworthy in this connection that the supposed repugnance felt by whites towards blacks did not prevent the former from employing the latter as servants in the most intimate settings.

No matter how imaginary the racial odour ascribed to blacks by whites might have been, it none the less had a profound effect both on how blacks were perceived by others and on how they perceived themselves. If you are told often enough that you have a foul odour, you come to believe it. Many blacks, repelled by their olfactory image, turned to perfumes and deodorants.[22] Their use of these products, of course, could do little to dispel a prejudice which was fundamentally cultural in nature and not physical. Any perfumes used by blacks would simply tend to emphasize their status as 'smelly', just as the perfumes used by the working classes in England were said to be an indication of their 'coarse' tastes.

Indeed, having a noticeable odour of any sort (with exception made in the case of women, who are allowed on occasion to wear perfumes) has traditionally been considered somewhat disreputable in the twentieth-century West, and particularly in the United States. Thus, according to Verrill: 'The Latins may reek of perfumes of Araby or elsewhere; the Englishmen may use conservative scents that hint of heather, of Castile or of mint, but not the dyed in the wool American.'[23] American nationalism here defines itself as staunchly inodorate as compared with the suspect scents of foreigners.

This olfactory social scale is the reverse of what it was in earlier ages in the West, when the better one's perfume, the higher one's social status. Important persons made powerful olfactory impressions. In 1709, for example, a French perfumer proposed that the different classes be scented differently. According to his scheme, there would be a royal perfume for the aristocracy, a bourgeois perfume for the middle classes, but only a disinfectant for the poor.[24] Now, however, power resides not with perfumed potentates, but with inodorate businessmen.

It often comes as a surprise to white Westerners, intent on condemning the odours of others, to learn that their own odour is not always pleasing to other peoples. In his examination of racial prejudice, the sociologist Robert Park tells the following story:

> A few years ago a Hindu acquaintance of mine, in explaining the opposition of his family to his marriage to an American woman, confessed that his father had written him saying he hoped, if no other considerations were sufficient, that the smell of an Anglo-Saxon would be sufficient to prohibit such a misalliance.[25]

The writer W. H. Hudson relates another anecdote of such olfactory antipathies. The story concerns a British army doctor in Bombay. Wanting to make himself well known to the local Eng-

lish society, the doctor gave his Indian servant orders always to
come into the church where he attended Sunday service to call
him out to a (supposedly) urgent case. One day the doctor
decided to attend a political meeting in a distant part of town,
and asked his servant to come along as a guide. The evening
turned out to be oppressively hot, and after sitting in the hall,
densely packed with Indians, for half an hour or so, the doctor
rushed out.

> After taking a few deep breaths he exclaimed: 'What a relief
> to get out! In another ten minutes I should have collapsed.
> The smell!'
> To which his servant promptly replied: 'Ah, Sahib, *now* you
> will understand what I suffer every Sunday when I have to go
> right to the middle of the church to call you out!'[26]

At times the discovery that whites can be foul-smelling comes as
a revelation of their human status to native peoples, unsure of
how to classify these strange foreigners otherwise. For instance,
in the 1930s a pair of Australian prospectors, the Leahy brothers,
went searching for gold in the highlands of New Guinea. 'Nigs
pong woefully,' wrote one of the brothers in his diary, expressing
his opinion of the odour of the New Guineans. The natives, for
their part, were kept at a distance by the strange scent – 'like
nothing they had ever smelled before' – of the whites. Could
these foreigners be human?, they wondered. It was only after the
whites left and the New Guineans had a whiff of the Leahy
brothers' latrine pit that they reached the conclusion that the
foreigners were, indeed, human.[27]

Smell, therefore, can play a role in many different forms of
social classification. At times it is an actual smell which triggers
an experience of difference on the part of the perceiver. Often,
however, the odour of the other is not so much a real scent as a
feeling of dislike transposed into the olfactory domain. In either
case, smell provides a potent symbolic means for creating and
enforcing class and ethnic boundaries.

SMELL POLLUTION

Just as society is criss-crossed with symbolic and actual olfactory
boundaries, so is the urban environment. The different olfactory
spaces of the modern Western city are largely a product of zoning

laws. These laws regulate the kinds of construction and sorts of activity that may go on in different areas, and by so doing also regulate the distribution and circulation of smells. Three kinds of urban domain may be distinguished for purposes of discussion: the industrial, the public, and the private or personal (the home).[28] The industrial domain includes industrial parks, garbage dumps, sewage treatment plants and the like – areas from which the general public is usually excluded and has no real interest in visiting anyway. Permanently bad or repugnant smells are usually considered legitimate in such spaces – an unavoidable byproduct of the industrial process.

Public space includes residential, shopping and entertainment areas as well as parks. In such spaces, the regime is usually one of olfactory neutrality. There are exceptions – odours of food and spices often waft from restaurants and bakeries – however, smells which would be considered offensive are usually banned from such areas by municipal sanitary by-laws.

Smells of all sorts become legitimate again in private space, the space of the home. Any offensive odours which escaped the bounds of one's home, however, would soon bring complaints from the neighbours, just as would a radio turned up too loud.

While it is possible to liken odour pollution to noise pollution, the analogy is a tenuous one, for smells are far more difficult to measure than sounds. A sound of a certain number of decibels is known to be harmful to one's hearing, but a smell of a certain concentration may not affect all people in the same way. Furthermore, smells which are tolerated in one setting will not be tolerated in another. Smell pollution is much more ill-defined than noise pollution.[29]

By way of illustration, the smell of manure, which seems natural in a rural setting, becomes intolerable in an urban one. In the countryside it signifies growth, in the city it would only mean decay. Certain foul smells may signify growth (if not of an organic sort) even in an urban setting, however. To the inhabitants of a mill town, for instance, the sulphurous reek of a pulp-and-paper mill might well mean money and progress. Furthermore, the inhabitants will probably have become habituated to the smell because of its constancy and will not therefore have any pronounced awareness of it.

Industry and farming, however, do create concentrations of foul odour not usually found in nature. In livestock farming, for example, animals are kept in much closer quarters than those in which they would live in a natural setting, and the odours produced by this crowding consequently reach an unnatural intensity. Even persons who don't mind a whiff of manure can find the smell of the modern 'factory farm' breathtaking. A similar olfactory intensity occurs in slaughterhouses in which immense numbers of animals are killed and dismembered in one building.[30]

The smell is tolerable for those who are accustomed to it and have a profit to make out of these malodorous businesses. For those who are not and do not, however, it can be unbearable. This conflict of interests often arises in areas which are both industrial or agricultural and residential. Slaughterhouses, for example, are sometimes located in semi-residential areas. The stench of the slaughterhouse is then often perceived as a form of smell pollution by the residents, as in the following report of a Montreal family living next to a chicken 'processing' plant.

> During the hot spell early this month, Felix Poser made the mistake of taking his cocker spaniel, Goldie, for a walk through the lane. By the time they got home, the dog stank of chicken and needed a bath.
> 'She smelled like something dead,' said Ann Poser, who sometimes takes clothes off the line and throws them back into the washing machine to get rid of the fowl [sic] odour.[31]

The plant owner admitted that the smell was 'something fierce', but still argued that he was complying with government health regulations and that a foul scent came with the territory: 'It's a meat plant, not a perfume factory.'[32]

When such cases make it to court, the traditional olfactory status quo is usually maintained. Thus, while a malodorous industry cannot set up in a previously inodorate neighbourhood, residents cannot demand that a malodorous industry which has traditionally operated in the area (and which complies with government regulations) cleans up its act. It is difficult to argue, as noted above, that foul odours are a hazard to one's health. Furthermore, it is often a question of applying vague laws to a poorly understood problem.

For smell-polluting companies, however, deodorization is not the only option. Another is to mask the unwanted odour with a pleasant neutral smell, for example that of bubble gum. One strategy explored recently has been to develop smell-suppressing sprays. These sprays contain odourless molecules called 'antagonists' that block one's ability to perceive certain odours. 'More important than creating new perfumes', says a researcher working in the field, 'is to develop antagonists which you could spray all over the men's room in Penn Station – and do a great service to humanity.'[33] Such olfactory screens obviate the need to clean things up. Bad odours simply aren't noticed any more. Imagine if this technique were carried over into the other senses and eyesores were made invisible or the noises of construction and traffic inaudible! Half of the sensory reality of our cities would vanish. Apart from the likely problems of safety, we would lose our ability to experience the environment we live in and to react to it according to that experience. The issue of smell pollution has moved beyond questions of what constitutes a bad smell and the importance of unfouled air; what is now at issue is whether or not we have a right to 'freedom of smell', or whether our olfactory environment can be censored with impunity by those who prefer to keep their stinks hidden.

THE STENCH OF AUSCHWITZ

It is not surprising that the sanitary reform movement should sometimes have been linked to movements for social reform, or that the elimination of undesirable odours should at times have been associated with the elimination of undesirable people. As we have seen in this and previous chapters, odour and morality have a long-standing association in the West, so that at times little distinction seemed to be made between physical stench and moral corruption. A good example of this is the statement made by an eighteenth-century sanitary reformer that prostitutes disappeared along with foul odours after the streets of Florence were cleaned.[34]

The drive to purify the Western social body of 'corrupt elements' reached its height in Nazi Germany. Jews, in particular, were characterized by the National Socialists as 'germ-carriers' and 'agents of racial pollution'.[35] Hitler wrote in *Mein Kampf*:

The cleanliness of [Jews], moral and otherwise, I must say, is a point in itself. By their very exterior you could tell that these were no lovers of water, and, to your distress, you often knew

it with your eyes closed. Later I often grew sick to my stomach
from the smell of these caftan-wearers . . .

All this could scarcely be called very attractive; but it
became positively repulsive when, in addition to their physical
uncleanliness, you discovered the moral stains on this 'chosen
people'.[36]

Through his racist politics of smell, Hitler sought to mark Jews
as undesirable and socially dangerous by projecting a foul odour
onto them and associating this odour with physical and moral
corruption.

The Hygienic Institutes set up in Germany during the Nazi era
had as their responsibility – along with the control of epidemics
and the study of bacteria – the distribution of the deadly gas
used to 'eliminate' Jews and others at Auschwitz.[37] In the words
of Deputy Party Leader Rudolf Hess: 'National Socialism is
nothing but applied biology.'[38]

The olfactory conditions in Auschwitz and other concentration
camps approximated those associated with factory farms and
slaughterhouses. On the route to the camp, prisoners were packed
tightly into cattle cars, forced to endure the odour of their own
waste and even their own dead. Within the camp, overcrowded
barracks and cells emitted a suffocating stench. Without the
means to keep themselves clean, prisoners lived in a state of
perpetual filth. A group of female internees is thus described, in
the words of one of them, as a 'herd of dirty, evil-smelling
women'.[39] The malodour of the prisoners confirmed their identification to their guards (and, at times, to themselves)[40] as 'stinking
Jews' and 'human filth'.

Worst of all was the persistent odour of burning bodies from
the crematoria. This was also the most troublesome from the
point of view of the Nazis, for the stench emitted by the crematoria invaded the domain of nearby homes. Rudolf Höss, the com-
mandant of Auschwitz, writes that:

During bad weather or when a strong wind was blowing, the
stench of burning flesh was carried for many miles and caused
the whole neighbourhood to talk about the burning of Jews,
despite official counter-propaganda.[41]

The Nazis were able to keep the sight and sound of their atrocities
within impenetrable walls, but smell escaped and breathed the

horrible truth. Nazi officials set themselves to work to solve the problem, but without success: the stench persisted. As Hans Rindisbacher writes, this obsession with the malodour of their deeds was perhaps 'the last remnant of bad conscience [of] the Nazi death machine'.[42]

However horrid the stench, the Nazis in charge of the camps had to learn how to live with it. The process of adaptation that took place among officials assigned to these camps is described by a Nazi doctor as follows:

I think I can give you a kind of impression of it. When you have gone into a slaughterhouse where animals are being slaughtered, ... the smell is also a part of it, ... not just the fact that [the cattle] fall over [dead] and so forth. A steak will probably not taste good to us afterward. [But] when you do that every day for two weeks, then your steak again tastes as good as before.[43]

In a strange metamorphosis of the traditional odour of sanctity and stench of sin dichotomy, Nazi officials argued that it was better for prisoners to '[go] to heaven in [a cloud] of gas' than to '[die] in shit'.[44] In other words, given that their death was inevitable, it was more humane to kill prisoners quickly with gas than to let them die slowly in squalid conditions.

Perfumes provided a striking contrast with the malodour of the camp. A former prisoner at Auschwitz describes the sensation the arrival of a scented letter made among her fellow inmates:

Of course everyone had to smell the perfume, so the note was passed around and sniffed ecstatically ... I pressed it to my nose and inhaled greedily.[45]

Certain Nazi officials were known for their use of perfume. One survivor remembers the infamous Nazi doctor Josef Mengele as 'smelling of eau de cologne' and being 'very sensitive about bad smells'.[46] Irma Griese, a Nazi officer known as 'the blonde angel', is also remembered for her scents:

Wherever she went she brought the scent of rare perfume. Her hair was sprayed with a complete range of tantalizing odors: sometimes she blended her own concoctions. Her immodest use of perfume was perhaps the supreme refinement of her cruelty. The internees, who had fallen to a state of physical degradation, inhaled these fragrances joyfully. By contrast, when she left us and the stale, sickening odor of human flesh, which covered the camp like a blanket, crept over us again, the atmosphere became even more unbearable.[47]

For the prisoners 'the blonde angel' is a perfumed *femme fatale*, torturing her captives with her scent. Griese herself, however, may well not have thought of her use of perfume as a tool of torture, but rather as a means of dissociating herself from the foul odours of the camp and maintaining a separate olfactory identity.

When the defeat of the Nazi forces became certain, Nazi officers busied themselves with removing the signs of their mass murders – dynamiting the crematoria, sweeping up stray bones and body parts – but the stench lingered on to confront the noses of the liberators. Here is a statement by one of them, presented in the form of a poem by Barbara Hyett:

The ovens,
the stench,
I couldn't repeat
the stench. You
have to breathe.
You can wipe out
what you don't want
to see. Close your
eyes. You don't want
to hear, don't want
to taste. You can
block out all the senses
except smell.[48]

THE FRAGRANT UTOPIA

In twentieth-century Western culture, the ideal society is presented as deodorized. Indeed, the fantasy worlds created for us by Hollywood on film are totally inodorate, existing only in the sensory domains of sight and hearing. These scentless representations, which continually produce and reproduce the world for us, reinforce the social drive for deodorization.[49]

In the medium of literature, the ideal of olfactory purity has been explored in the utopian visions of several modern authors.

In 1911, the German writer and philosopher Mynona (Salomon Friedländer) published a short story entitled 'On the Bliss of Crossing Bridges'. In this story a German scientist, Dr Krendelen, invents a chemical formula to purify the planetary atmosphere of bad air: 'For bad air is the misfortune of mankind... The improvement of the air is the surest way to improve humanity, better than all philosophical moralizing!' The scientist realizes that only a few persons will be able to survive in the rarefied atmosphere; none the less he resolves to go ahead with his plan for world purification. Almost at once people begin to die. Their bodies, however, burn 'without a trace of corrupting odor in the delightful air of early spring'.[50]

At last the purification is complete. 'Nothing was left of corruption. Victoriously it was all banished and masked by the scents of fresh purity that now virtually exploded!' Death is dispelled along with stench and, with no foul odours to remind them of how things used to be, the past is promptly forgotten by all, 'so that [Dr Krendelen] did not even become famous!'[51]

A novel with a similar theme, *The New Pleasure*, was published by John Gloag in 1933.[52] In this novel, an English scientist discovers a substance, named voe, which renders the human sense of smell hypersensitive. Industry at once realizes the commercial potential of this new product and begins to market it. With voe, humans are able to enjoy as never before attractive aromas.

At the same time, however, the new olfactory sensitivity makes foul smells unbearable. As a result, an olfactory revolution takes place. No one wants to work in malodoured trades any more. Pollution and poor sanitation are no longer tolerated and people flee the stench of the cities for the countryside. In the United States, agitation resulting from interracial olfactory antipathies almost causes a second Civil War.

After an initial period of disorder, measures are taken to eradicate foul odours from the environment. This eventually leads to new and improved forms of agriculture, architecture, town planning, industrial development, transportation and even morality. Cities become fragrant parklands with superior sewage systems. After his death, the discoverer of voe, Professor Frankby, is made a saint for his invaluable contribution to human welfare.

Both Mynona and Gloag imagine a world purified of stench. In Mynona's case, however, there is a dark edge to his vision: olfactory purity comes at the cost of human life and morality.

Gloag's utopia is more positive: a heightened sensitivity to odour results in beneficial social and technological reforms. Furthermore, while in Mynona's story the chemical invention acts directly to purify the air, leaving humans to adapt or die, in Gloag's it instead serves to refine the sense of smell, leading humans to deodorize their environment. These different perspectives are also reflected in the fate of the two different protagonists. Dr Krendelen becomes anonymous, his revolutionary invention forgotten with the corrupt air it purified. Professor Frankby, on the the other hand, invested with a sweet odour of sanctity, is canonized by a grateful humanity.

These two stories invite interpretation within a sociohistorical context. Mynona's tale cannot but be seen as a foreshadowing of the Nazi quest for social purity. The chemical formula used by Krendelen to rid the air of undesirable odours parallels the chemical gases used by the Nazis to rid society of 'undesirable' persons. In Mynona's story, however, the bodies of the people killed by the chemical burn without a trace of odour, while the bodies of the Nazis' victims, as we have seen, emitted an unforgettable odour. The Nazis did not have Krendelen's ability to wipe away the traces of their deeds (or 'antagonists' to render the stench imperceptible). Nor, however, did they succeed in implementing their 'final solution', as he did.

Gloag's novel, on the other hand, reads like an allegory of the modern movement of sanitary reform: a growing intolerance of foul odours leads to major improvements in hygiene. Due to the influence of voe, 'planning new sewers, abolishing cesspools, purifying water supplies' and other sanitary improvements become the most pressing concern on the urban agenda.[53] The difference is that, while in Gloag's account it is a particular substance which renders humans hypersensitive to stench, in actuality this hypersensitivity is cultural in origin, occurring as a result of certain social trends. The consequence, in either case, is that dreamt of by the sanitary reformers of the nineteenth century: the widespread deodorization of the human environment.

Perhaps the best-known example of a fragrant utopia is that described by Aldous Huxley in *Brave New World*. 'Civilization is sterilization' is the motto of Huxley's utopia, with sterilization applying not just to dirt, but to humans (babies are created in test tubes) and human emotions (too 'messy' for an ordered society). Workers are kept happy and productive through equal parts of sensory conditioning and gratification. Throughout the novel, the manipulation of odour stands as a sign for the artificially controlled and enhanced environment of the utopia. Scents of rose and asafetida help condition children to accept their prescribed roles in life. Bathroom taps dispense not only water but eight different perfumes. Olfactory entertainment is provided by a 'scent organ':

> The scent organ was playing a delightfully refreshing Herbal Capriccio – rippling arpeggios of thyme and lavender, of rosemary, basil, myrtle, tarragon; a series of daring modulations through the spice keys into ambergris; and a slow return through sandalwood, camphor, cedar and new-mown hay (with occasional subtle touches of discord – a whiff of kidney pudding, the faintest suspicion of pig dung) back to the simple aromatics with which the piece began.[54]

An Indian reservation in New Mexico appears as the antithesis of utopian civilization. The Indians are perceived by the utopians as dirty and foul-smelling, bound by the disgusting organic processes of life and death. Significantly, while the primitive conditions of the reservation are being described to a pair of utopians going there for a visit, one of them remembers that he left the eau de cologne tap running back in his hotel room. This utopian flow of fragrance parallels the flow of filth in the reservation.[55]

One of the 'savages' leaves the reservation for civilization. There he becomes enamoured of a utopian woman through her perfume:

> Opening a box, he spilt a cloud of scented powder . . . Delicious perfume! He shut his eyes; he rubbed his cheek against his own powdered arm . . . scent in his nostrils of musky dust – her real presence. 'Lenina,' he whispered. 'Lenina!'[56]

Soon, however, he comes to the conclusion that the woman's sweet scent is only an artificial lure, masking her true stench of immorality. The 'savage' then rejects the false delights of civilization and returns to the 'primitive' customs of the reservation, whipping himself in penance until his blood runs. Even this act, however, is perceived as entertainment by the utopians, and the 'savage' ends up hanging himself.[57]

In *Brave New World*, therefore, fragrance stands for artificial and superficial pleasure, while foulness stands for unpleasant but meaningful reality. There is no compromise between the fragrance of civilization and the stink of savagery. The former repels through its moral emptiness, the latter through its aesthetic ugliness. Once society is (metaphorically as well as actually) deodorized and perfumed, Huxley seems to be arguing, the only alternative to it is the frank squalor of stench – the ultimate gesture of defiance. In the end, Huxley leaves the reader with the feeling that, all things considered, the odour of flowing blood is of more value than the fragrance of perfume flowing from a scent tap.

Chapter 6

The aroma of the commodity
The commercialization of smell

In an essay on 'Consumer Culture and the Aura of the Commodity', Alan Tomlinson notes how in today's society commodities have acquired an 'aura', an air of fantasy, which goes beyond any practical purpose they may serve.[1] Items such as jeans or watches are bought as much or more for the images and lifestyles they project through advertising as for their practical usefulness. In effect, the aura – the shimmer of meanings and associations – surrounding the commodity has eclipsed the commodity itself in terms of importance in the marketplace. We acquire items for their style, more than for their function.

Nowhere is this phenomenon of 'image marketing' more apparent than in the advertising of deodorants and perfumes. The control of body odour is a major preoccupation of Westerners, who have made the deodorant and toiletry industry into a billion dollar business. Although natural body odour is stigmatized and suppressed, artificial body odour – in the form of perfumes and colognes – is condoned and even celebrated. Thus, while deodorants strip the body of its natural olfactory signs, perfumes invest it with a new, 'ideal' olfactory identity. These ideal identities are promoted by the 'dream merchants' of the perfume industry who assure consumers that all good things come to those with the right scent.

Such techniques of olfactory management are not limited to toiletry products. Fragrances added to products such as detergents and house paints give a wide range of commodities an olfactory aura. These added aromas carry meanings of status, of freshness, of effectiveness, without in any way being necessary to those products' actual performance. This 'aroma of the commodity', which often works at a subconscious or barely conscious level, has nevertheless been shown by market research to be crucial to the prospective consumer's perception of a product as desirable and worth buying.

Artificial essences are also present in much of the food we eat, determining how our meals smell and taste. These synthetic flavours are the creations of biochemists who work in laboratories composing imitation savours of everything from strawberries to pizza. In the premodern West, the quintessential meal was distinguished by the artifice with which it was prepared – pastries garnished with candied flowers, marzipan shaped to look like lions and peacocks, and so on. In the contemporary West, meals are distinguished by the artifice of their flavours.

The techniques of olfactory marketing discussed here are those current in the urban centres of the First World. Modern, mass-produced fragrances and flavours are, however, sold in most of the Third World as well, where they earn large profits for their manufacturers. Poverty, apparently, is no barrier to the massive consumption of such goods as perfumes and artificially flavoured foods. As an executive from International Flavors and Fragrances Inc. noted: 'the poorer the malnutritioned are, the more likely they are to spend a disproportionate amount of whatever they have . . . on some simple luxury.'[2] When real prosperity is out of reach, the poor must make do with a whiff of prosperity instead, it seems.

The means by which fragrances are marketed in Third World regions often present interesting and ingenious combinations of modern and traditional practices. In the Amazonian region of Brazil, for example, local 'Avon ladies' sell their wares from door to door, hut to hut, in exchange for regional products.[3] In North America, Avon's selling style is considered passé – North American women are not interested in selling door-to-door, and even when they are, there are not enough women who stay at home to answer the door any more. In Brazil, home sales still work, and there are 60,000 Avon ladies working the Amazon alone. In almost inaccessible mining outposts, men and women eagerly buy up colognes with names such as 'Crystal Splash', and 'Charisma' – each bottle costing a gram of gold. In rural villages, reached by canoe, 'two dozen eggs buys a Bart Simpson roll-on deodorant; 20 pounds of flour gets you a bottle of cologne.'[4] Thus, the aroma of the commodity has penetrated even into the Amazonian rainforest, though the techniques by which it is marketed differ somewhat from those of the industrial West.

Interestingly, despite their enormous commercial value, product scents do not technically constitute property under the law. This makes it possible for competitors to reproduce the aroma of a successful commodity in an imitation product. As a commodity's consumer appeal is often conveyed by its particular scent, this involves not simply duplicating an odour, but also appropriating the whole constellation of perceived values associated with it. Not surprisingly, marketers wish to prevent such appropriation of their product fragrances by establishing an exclusive property right to them (as they currently do with logos and other visual insignia). However, as we shall see below in 'Trademark Scents', the question arises of how an odour, an air, can be the subject of property rights. The conflict lies between fragrances as the stuff of dreams, and fragrances as the essence of material culture.

We close this chapter, and the book, with a discussion of smell and postmodernity. If sight – panoramic, analytic and linear – is the sense of modernity, is smell – personal, intuitive and multidirectional – the sense of postmodernity?

BODY ODOUR

Concern with body odour and with methods of suppressing it has existed in the West since antiquity. What is new in our era – the era of consumer capitalism – is the availability of ready-made, mass-produced products to deal with body odours and the advertising used to promote them. Furthermore, due in part to these new techniques of production and marketing and in part to lifestyle changes, whereas in previous centuries it was largely the well-to-do who were preoccupied with 'smelling sweet', this concern has now penetrated the consciousness of all social classes.

This novel form of capitalist penetration may be traced in part to certain developments in advertising tactics in the 1920s. Consider the case of 'Listerine'. Listerine had been sold as a general antiseptic for home and hospital use since the 1870s. In 1920, however, it was reinvented as a mouthwash. The new Listerine ads were modelled after the 'advice to the lovelorn' columns which had proved so popular in the tabloids. In one advertisement, the picture of a young woman peering questioningly into a mirror introduces a story entitled 'What secret is your mirror holding back?' The accompanying text makes note of all of the girl's 'advantages': she is not only beautiful but talented, educated and better dressed than most. However, in the one pursuit that matters, the girl remains a failure:

> She was often a bridesmaid but never a bride. And the secret her mirror held back concerned a thing she least suspected – a thing people simply will not tell you to your face.

> That's the insidious thing about halitosis (unpleasant breath). You, yourself, rarely know when you have it. And even your closest friends won't tell you.[5]

This ad is far more than a business announcement: it is a 'socio-drama', with all the ingredients of a tragedy.[6] The ad invites the reader to identify with the protagonist and to suffer vicariously her unhappy plight. Identification is enhanced by the device of referring to the mirror as 'your mirror' in the title, and oscillating between 'she' and 'you' in the text. The unhappy girl in the ad in this way becomes the reader's alter ego.

The ad succeeds in building up tension in the reader, and not a little paranoia. As the advertisers had realized, body odour is a perfect subject for a marketing campaign based on nameless fears. Individuals are unaware of their own smell, it cannot be 'seen' (as one's visual appearance can) in a mirror, and politeness decrees that it should not be broached by 'even your closest friends'. It is only through the ad, which speaks with the voice of an 'objective' third party, that one can be openly warned of the dangers of body odour.

At the same time as the ad makes the reader aware of the devastating social consequences of body odour, though, it holds out the promise of relief, of catharsis: if the troubled girl in the ad is ignorant of the nature of her social shortcoming, the reader is in the know, and only has to go and buy the product to avoid the same fate. The product promises to shield its buyers from any further social shame, befriending them in a way no one else will. The tactic worked: the profits of the manufacturer of Lister-ine, Lambert Pharmaceutical Company, went from $100,000 per year in 1920 to over $4 million in 1927.[7] There was no change in the substance of Listerine, only in its associations.

The use of the term 'halitosis' represented another major breakthrough in advertising practice. Giving bad breath a new name enabled the advertisers to talk about the very thing no one (supposedly) talked about. The scientific sound of this term, which was in fact exhumed from an old medical bulletin, also helped: it made bad breath sound like a medical condition. As a medical condition, bad breath became something which could and should be treated.

The medicalization of bad breath proved an extremely effective ploy, as evidenced by the way 'the halitosis style' or 'the halitosis appeal' – as this new advertising strategy came to be known – was emulated by others:

> In unmistakable tribute, [advertising] copywriters soon dis-
> covered and labeled over a hundred new diseases, including
> such transparent imitations as 'bromodosis' (sweaty foot
> odors), 'homotosis' (lack of attractive home furnishings), and
> 'acidosis' (sour stomach) and such inventive afflictions as
> 'office hips', 'ashtray breath', and 'accelerator toe'. Needless
> to say, most of these new diseases had escaped the notice of
> the medical profession.[8]

Advertisers, while ostensibly empathizing with the fears and anxieties of the general public, were mainly interested in capitalizing on those fears – the better to reduce buyer resistance to their products. Thus, for example, the anxieties of the average citizen concerning job security became one of the standard themes of the advertising dramas of the Depression years.

> Listerine tied mouthwash to depression fears with a January
> 1931 ad entitled 'Fired – and for a reason he never suspected',
> a theme that Lifebuoy Soap employed several months later in
> 'Don't risk *your* job by offending with B.O.' 'Take no chances!'
> warned Lifebuoy. 'When business is slack, employers become
> more critical. Sometimes very little may turn the scales against
> us.'[9]

Through texts like these, the advertisers not only articulated and gave definition to otherwise diffuse fears, they also humanized the impersonal forces of the marketplace and personalized the crises in the capitalist system that were responsible for people losing their jobs. Instead of perceiving the system to be at fault for their economic plight, the unemployed would pin the blame on their own persons, specifically on the odours emanating from their bodies. As a result of this displacement of blame, people

might be able to make sense of their experience of the marketplace, though only at the cost of coming to feel alienated from
their bodies.

The role of advertisers as the apostles and apologists of
modernity – that is, as interpreters and rationalizers of the
ever-changing conditions of life in the twentieth century – is
particularly apparent in the case of the massive advertising campaign sponsored by the Cleanliness Institute of the Association of
American Soap and Glycerine Producers. Many of the Cleanliness
Institute ads were constructed around what Roland Marchand has
called the 'parable of the First Impression'. In one ad depicting a
job interview, a business executive stares sternly across his desk
at a man with a troubled expression. The latter turns in his
chair and gestures at a huge spectre of himself, cringing with
embarrassment and self-doubt, that looms over his shoulder. The
interview is an apparent failure, and the reason (in the words of
the copywriter) is that the applicant:

> was his own worst enemy. His appearance was against him and
> he knew it. Oh why had he neglected the bath that morning,
> the shave, the change of linen? Under the other fellow's gaze
> it was hard to forget that cheap feeling.
>
> There's self-respect in soap and water. The clean-cut chap
> can look any man in the face and tell him the facts – for when
> you're clean your appearance fights *for* you.[10]

The social context of this parable was that there was growing
uncertainty as to whether a person of ability and character would
necessarily win out in the scramble for jobs and for success. The
sense had emerged that getting a job and getting ahead in life
depended more on making the right first impression than on any
intrinsic qualities of the person. This sense was inspired and
confirmed by the growing anonymity of business and social
relationships, as well as the faster pace of 'modern life'. 'Quick
decisions' had become the norm, particularly, it seemed, with
regard to hiring.[11]

One of the more striking associations the Cleanliness Institute
ad brings to light is the peculiarly modern association between
inodorateness and power. In contemporary urban life, the strong
man is neither the sweaty labourer nor the perfumed aristocrat,
but the inodorate, clean-cut businessman. By removing dirt and
body odour, soap, the ad tells us, confers both social equality

and the power of objectivity on its users: 'The clean-cut chap can look any man in the face and tell him the facts.'

A comparison of the approaches of the Listerine ad featuring a woman with the soap ads directed at men shows that each one is aimed at exploiting what is considered to be the pre-eminent concern of the sex in question: in the case of women, getting a man; and in the case of men, getting a job. Women must take care not to offend potential husbands with their odour, and men must take care not to similarly offend employers. In either case, the emphasis is not on internal worth but on external impression. One's value is measured by the approval or rejection of others.

Summing up, we may say that the effect of the Listerine and other ads for deodorant products was to open a gap between self and body, and to insinuate the product being promoted into that gap. The message of the ads is that the product could protect the self from the social disaffection and disgrace to which the body might otherwise expose one by emitting antisocial odours. The desired result of the ads is for people to feel alienated from their bodies and dependent on toiletry products to save them from themselves.

Once deodorants have 'saved' us from social rejection, the question arises of how to win social approval. This is where perfumes, with their reputed image-enhancing powers, come in.

PERFUME

Advertising perfumes presents more of a challenge than advertising other goods, for scents are notoriously difficult to describe or evoke through language. There are no adequate words to describe the character of most perfumes, which poses serious problems for marketing. The solution has been to simply avoid giving product information and instead focus on evoking fantasies. Let us compare two ads for musk perfume. In one from 1857, for Harrison's *Musk Cologne*, there is a drawing of two musk deer; in a 1986 ad for Coty *Wild Musk*, the image is of a woman pressing a man against the wall, tearing the shirt off his torso. The first presents us with information about the product – musk comes from musk deer; the second presents us instead with the presumed effect of the product – heightened sexual attraction. There has been a shift, from origin or cause to effect. It would seem that the effect is all that interests us any longer.[12]

There is a sense in which advertisers have turned the very non-discursivity of perfume into a sales asset. As Tom Zelman observes in 'Language and Perfume':

> Unable to find discursive symbols [i.e. words] to represent a scent, the advertiser instead claims that the scent itself is suggestive of sexuality, wealth, rugged individualism, and so on. Copywriting then becomes dedicated to the task of creating connotations for a particular indescribable scent to give symbolic import to the fragrance.[13]

Perfumes have therefore been advertised less through words than through images, making perfume ads the forerunners of modern image-based advertising.

No perfume ad would show us a perfume being concocted in a laboratory by a white-coated technician. While such 'scientific' imagery might work for some products, such as face cream, which gain respect by being invested with a scientific or medical authority, it would detract from the primary selling mythos of perfume. That mythos is that perfume has – or rather, *is* – an inexplicable, pseudo-magical force. Furthermore, laboratories are conceived of as sterile, inodorate places, an image which contradicts the fertile redolence promised by perfume. Perfume ads, therefore, present us with seductive imagery of beauty, wealth, exoticism, love and sexual power in order to create irresistible associations with a brand of scent.

For example, *Musk* by English Leather was advertised as 'the missing link between animal and man', thus promising to reinvest men with the animal (i.e. sexual) power they have lost by becoming 'civilized'. The perfume *20 Carats*, on the other hand, made blatant (but ultimately less successful) pitch to consumers' desire for wealth in its advertising motto: 'Smell rich.' *Rive Gauche* by Yves Saint Laurent has relied on exotic appeal for its selling power: 'It has a spirit of a Parisian café, alive with wine, laughter, and love.' Again, the language is used to create a conception of the product, not to describe the product itself. Zelman notes:

> Certain ads play up the very ineffability of scent: Unable to describe an odor, they call it 'mysterious' and the inscrutability of a scent that defies language is transferred onto the wearer, or so it is claimed. *Infini*, for example, simply carries the tag: 'Because I like to be mysterious.'[14]

Thus, the intrinsic 'mystery', or indescribability of odour as a medium, is transferred to the wearer.

Perfumes have also been represented as bypassing language, or speaking for themselves. Hence the *Brut* ad campaign in which baseball star Henry Aaron endorses the product by asserting: 'When I'm off the field, I let my Brut do the talking.' This claim would definitely have an appeal for men or women who feel shy or awkward about communicating with members of the opposite sex. There is no longer any need to express one's feelings in words, one's scent will say it all – and bring about the desired effect. 'Like a graceful gesture,' reads an advertisement for *Eau du Soir*, 'Eau du Soir reveals unspoken hopes and dreams.'

The final step in this progression of advertising techniques is to go beyond words altogether – except, of course, for the essential brand name. This strategy is apparent in the ads for *Obsession* by Calvin Klein, which feature shadowy, nude figures entwined with each other in a gauzy blue light. The situation depicted in the photograph is far from clear, and there is no explanatory text. The consumer is left to interact directly with the sensual imagery of the photograph without the imposition of language. The advertising technique is no longer to convince the reader of the value of the product through words, but rather to present an arresting image which will elicit an emotional reaction from the consumer. Such ads are suggestive only. They package 'experiences' which seem to go beyond words and so have to be filled in, completed by the consumer. It is 'mood advertising' – the very antithesis of reasoned argument.[15]

Modern perfume advertising has also been greatly aided by the invention of scent strips, microscopic capsules of scent attached to a strip of paper which will release their odour when scratched or broken. This invention has enabled perfumes to advertise themselves directly, through smell, instead of having to rely on words or pictures. Nevertheless, visual imagery continues to play a central role in perfume marketing. In 'Focus on Fragrance: Photography in Perfume Advertising', Fred Naraschkewitz writes: 'Photography's task in perfume advertising is to translate into visual terms the olfactory impression of a perfume, together with the lifestyle associated with the fragrance by the marketing experts.'[16] In our society, at least, smell is too underdeveloped a sense for odours to speak for themselves, despite the claims of perfume ads.

Indeed, it is a curious fact that few perfume ads actually refer to smell, the sense to which their product is surely directed. Instead, advertisements speak in terms of 'enchantment', 'sensuality' and 'mystery'. Sometimes more attention is given to the bottle which contains the perfume than to the scent itself. This reflects the low and ambiguous status of smell in our culture. In an attempt to capitalize on the prestige and power of other senses, perfumes have been created with such non-olfactory names as *Jazz*, *Echo*, *Touch*, *Farhenheit* and *Photo*. This last, a men's cologne, presents itself as the olfactory equivalent of *Playboy*. The bottle is fashioned so as to represent a camera, with its cap simulating a lens cap. The image used to advertise the scent is that of a man photographing a beautiful, naked woman. Male voyeurism is thus transformed into perfume.

The topic which begs discussion here is the gender division of perfumes.[17] Perfume has traditionally been considered a woman's product in twentieth-century North America. The great majority of perfumes and colognes are therefore created for and directed to women. Perfume ads over the last decades reflect the changing roles and images of women in the modern West.[18] In the 1950s femininity, elegance and charm were the themes of perfume advertisements. In the ads we are invited to participate in an evening out, a high society ball, a night at the opera, evoked by the evening gown, the pearls, the marble staircase, gold mirrors, candelabras. The slogan which epitomized the role of women during this period was 'To be a real woman is to bring out the best in a man.' Fragrances are worn by women for the pleasure and enticement of men. This is reflected in the way the male figure, if present, looks on – often from a position above the woman, while she gives him her best come-hither look. This form of advertising continues today, for example in the ads for *White Shoulders*.

In the 1960s and 70s perfume ads began to work with images of the 'sensuous woman' and the 'natural woman'. The former is a *femme fatale* who is willing to go to any lengths to capture her male prey. Her perfumes are musky or spicy. Perfumes which make use of *femme fatale* imagery include *Magie Noire*, *Poison* and *Opium*. The natural woman is a 'flower child' or a 'sportswoman', who rejects artificiality and uses light, fresh fragrances. Estée Lauder's *White Linen*, advertised as 'crisp, refreshing', is an example of the 'natural' approach.

In the 1970s *Charlie* made a breakthrough in perfume advertising by using the image of a 'liberated' woman to sell its product.[19] In a popular *Charlie* ad, a woman and a man, both carrying briefcases and both in business dress (he in a suit, she in a black skirt and polka dot jacket) are shown from the back. She is slightly taller and is reaching to pat his backside. 'She's very Charlie,' the ad proclaims. Here, the masculine name of the perfume, as well as the dominant role of the woman in the photograph, indicates that women who use it are usurping traditional male prerogatives. At the same time, this is being done in a playful way, as conveyed by the use of the nickname 'Charlie' (rather than Charles), and the mischievous pat on the male behind.

Finally, in the 1980s and 90s we see perfumes increasingly advertised with images which suggest self-fulfillment through perfume. In such ads, a woman will be shown holding or caressing an often enlarged perfume bottle. The picture is self-contained, no man appears to be necessary, the relationship is solely between a woman and her scent. In one such ad, for *Bijan*, the caption reads: 'mario, you might as well know the truth ... I'm in love with *bijan*.' Rather than being an element in the seduction of men, perfume is a source of solitary female pleasure.[20] Here the product does not simply evoke the fantasy, it *is* the fantasy.

The sale of fragrances for men, while still far behind that for women, has grown dramatically over the last decades. This is due largely to the relaxation of gender boundaries from the 1960s on. Just as child-rearing and home-making were no longer exclusively feminine domains, nor were colourful clothing or fragrance. The inodorate, dark-suited businessman might have been a force to reckon with in the world of finance, but he wasn't having much fun or satisfaction on a personal level. The time was ripe for expanding the male cultural horizon to include elements of play and fantasy.[21]

Despite the climate of change, during the 1960s and 70s perfumers still had to contend with a very strong taboo against the use of fragrance by men. Perfume remained so closely associated with women, that not even the word could be used with reference to scents for men, and terms such as cologne, eau de toilette and aftershave had to be used instead.[22] Furthermore, to compensate for their feminine associations, men's fragrances have had to be imbued with images of exaggerated masculinity. Names such as

Brut and *English Leather* were given to colognes for men in order to counter any suggestion of effeminacy. Male icons, such as cowboys and baseball players, were enlisted to promote colognes so as to reinforce their masculine nature.

One of the best selling pitches used to date was to convince the consumer that, far from being feminizing, the touted scent in fact contained the essence of masculinity, so that men who wore it would become irresistible to women. The well-known ad in which a mild-mannered man becomes so attractive to women by using *Hai Karate* cologne that he has to fend them off with karate is a successful example of this approach. A more recent example is the ad for Bijan's *DNA* for men – sold in a bottle shaped like a strand of DNA – which suggests that the fragrance somehow contains the very building blocks of masculinity. (Although the ad states cryptically in a note that '*DNA* fragrances do not contain deoxyribonucleic acid (DNA) except as included in the ingredient list on product packaging.') In such ads, perfume is shown not only to complement masculinity, but to actively increase it.

Advertisers also know, however, that women buy colognes to give to men as much as men buy them for themselves. At the same time as they appeal to men, therefore, ads for male fragrances (which often appear in women's magazines) promise women that these scents will make their men more handsome, more virile, and certainly more fragrant. Besides, colognes have filled a need for a handy all-occasion present to give to men, becoming an established gift for boyfriends, husbands and fathers.

Contemporary advertisements for men's fragrances are more sophisticated than their predecessors. Masculine traits are still emphasized and the word perfume is still taboo (*New West* fragrance for men advertises itself as a 'skinscent'). However, many male colognes now promise sensitivity as well as masculinity. *Jilsander* has as its slogan 'Feeling Man'. *Tsar* ads show an artistic-looking man gazing up dreamily at a sculpture. 'Vivre en *Tsar* est un art,' the slogan proclaims. In *Boss* ('This commanding men's fragrance lends an air of authority to men of all ages and backgrounds') and *Givenchy Gentleman* ('Think of it as investment spending') we even find an aroma for the businessman.

Increasingly popular for both men and women are signature scents, perfumes which carry the name of a well-known fashion designer. Examples include *Gucci*, *Versace* and *Sung*. These scents offer the status of their designer names as well as presenting themselves as designer accessories, an olfactory equivalent of a *Versace* scarf or *Gucci* shoes. By using the product one has the satisfaction of feeling one belongs to a cultural elite, an international community of wealth and discrimination.

With many current fragrances there is little to indicate whether they are meant for men or women. Advertisements may show only the perfume bottle, or else a photograph of a man and woman together, with no clue as to which, if either, is supposed to be sporting the scent in question. Even the scents themselves are sometimes difficult to classify as a traditional man's or woman's fragrance. Perhaps this is the first step towards unisex fragrances, scents which can be worn by either women or men, as they once could in the heyday of perfume in the seventeenth and eighteenth centuries – a trend of which 'Eau de Cologne' is the sole survivor.

THE FRAGRANT PRODUCT

Although the perfume business represents a substantial industry, most commercial fragrances are used to sell other sorts of products – from paints through detergents and stationery to cars. The practice of adding fragrance to products to increase their appeal has been around for some time. A perfumery book published in 1940, for instance, proclaims that:

> Even now, the lingerie shops display scented underwear; our steam laundries deliver our weekly wash delicately perfumed; there are scented paints and lacquers for finishing bedsteads, dressing tables and writing desks, and cigarettes are sprayed with perfume made up of rum, vanillin, geranium and other odors to impart a pleasant odor as they are smoked.[23]

What was at first something of a novelty, however, has been found to make good business sense. Fragrance marketers like to stress that, unlike the other senses, which convey messages to the brain through a series of intermediary synapses, smell has a 'direct connection' to the brain.[24] The parts of the brain smell connects with are, moreover, held to control memory, mood and emotions. Thus product fragrances theoretically work directly on consumers' emotions – a marketer's dream!

Whatever the connections between smell, the brain and

emotions, studies have shown that consumers do generally prefer scented products to unscented ones. Market tests have also revealed that, when the right scent is added to a product, that product will not only be perceived as more pleasing by the consumer, but as more effective and of superior quality. A classic example of this is the impression produced on consumers by the addition of a lemon scent to *Joy* detergent in 1966. Although the lemon scent in no way altered the substance of the detergent, it gave consumers, who associated 'lemonness' with grease-cutting ability, the sense that *Joy* had increased cleaning power.[25]

Such olfactory psychology works even when the added scent has no direct association with product performance. For instance, in one consumer test of shampoos, a shampoo which had been ranked last in general performance in an initial test, was ranked first in a new test after its fragrance had been changed: 'In the new test, the consumers stated that the shampoo, with the improved fragrance, was easier to rinse out, foamed better, and left the hair more lustrous and glossy.'[26]

Perhaps stemming from the original use of smell to tell whether something is good to eat or not, odour is an important means by which consumers judge the value and effectiveness of a product. Unlike packaging, which is external to the product, odour is perceived as intrinsic to it and therefore revelatory of its essential worth. Stephan Jellinek writes in *The Use of Fragrance in Consumer Products*:

> When a consumer uses fragrance as an indicator of a product's ability to deliver certain benefits ... an assumption is always present. That is, the consumer assumes that the fragrance is an organic, integral part of the product, inseparable from the product as a whole – just as the taste of an apple is an integral part of the fruit and can tell us a lot about it: what strain it is, how ripe it is, or how long ago it was picked.[27]

It is in marketers' interests, according to Jellinek, not to draw the consumers' attention to the fact that scent is an added feature of a product. This would destroy the illusion that the odour is an integral part of the product and the scent would come to be thought of as nothing more than a superficial characteristic. None the less, in certain cases the same product is sold with different fragrances, or produced in scented and unscented forms. With some products (for example, shaving cream), alerting the consumer to the fact that the scent is simply an added characteristic does not appear to harm how the product is perceived, because 'the performance of such products can be so readily evaluated that the user would not tend to employ fragrance as a signal.' 'Nevertheless,' Jellinek warns marketers:

> the practice is risky because people generalize. As they become aware that fragrance is a superficial, interchangeable attribute in shaving creams, antiperspirants, and hair lacquers, they are less likely to consider it as an indicator of essential product qualities in other categories.[28]

Once a marketer realizes that an added fragrance will boost the selling power of a product, the question becomes which fragrance to add. This depends to some extent on the message the marketer wishes the fragrance to carry. Contrary to what the scanty nature of our olfactory vocabularies might lead one to expect, a code of smell does exist in contemporary Western culture – that is, many odours are coded with meanings that are common to a large segment of the population. For example, the lemon scent added to *Joy* 'told' consumers that the detergent cut through grease. If one wanted to convey, rather, that a detergent was gentle on one's hands, a soft scent would be suitable. Jellinek notes that adding a mild fragrance to a detergent advertised as strong would serve to reassure users that it would not harm their skin.

Certain fragrances will convey several connotations, not all of which may be suitable for a given product. In one consumer test a pine fragrance evaluated as 'fresh' and 'clean' was added to facial tissues. When the tissues were then tested, however, they were considered harsh and rough. This was because the pine fragrance also carried associations of 'rough' and 'hard' – undesirable qualities for facial tissues. Similarly, candy scents rated pleasing and fresh do poorly when added to toothpaste as they also carry the meaning, 'bad for your teeth'.[29]

When adding a fragrance to a product, marketers must take into account gender differences in olfactory associations. A 'baby powder' scent might be pleasing to women, for example, but displeasing to men. Cultural differences must also be considered, particularly when a product will be sold in various countries. In Europe, *Sunlight* laundry soap sold well with an added scent of citronella. In North America, however, where citronella is associated with mosquito repellent, its presence in a laundry product is not considered appropriate.[30]

Furthermore, members of different generations will have different collective olfactory preferences. For example, persons born in the 1950s and later might enjoy fruit scents in soaps, while earlier generations will prefer floral fragrances. Jellinek (writing in 1975) notes that:

> Ten years ago a herbal scent in a shampoo, if it indicated anything at all to the average user, might well have communicated 'for people who believe in strange folk medicines'; today it means 'modern' and 'living in harmony with nature'.[31]

Just as specific market niches may be successfully targeted through the selection of a particular fragrance for a product, so may segments of the population be alienated. The product perfumer's task is a delicate one, therefore, which demands constant vigilance of the associations which a given fragrance may evoke in different people.

One traditional drawback concerning the use of odours to enhance product attractiveness has been that the scent does not act on consumers until they are in contact with the product. Scratch-and-sniff scent strips have now made it possible for odour to be advertised at a distance, just as photographs are used to advertise visual appearance at a distance. Apart from their use to publicize perfumes and colognes, however, such scent strips have up to now been employed mostly as an advertising 'gimmick'. For example, BEI Defense Systems Co. made use of a scent strip in an ad touting the power of its Flechette rocket system. The ad depicts an anti-aircraft rocket destroying an enemy helicopter. The slogan, 'The smell of victory', is accompanied by a scent strip which emits the odour of gunpowder.[32] Gunpowder, presumably, is one of those scents which appeal to army generals.

Another company, American Republic Insurance, concluding that mint was the scent of money, sent out mint-scented dollars in their direct-mail advertising.[33] This was something of a verbo-olfactory pun, as mint, of course, is the name for a place where money is coined. Perhaps the insurers hoped that consumers, subliminally affected by the associations of the mint smell, would think that their money would multiply in American Republic Insurance like coins in a mint.

Odours, in fact, are increasingly being promoted as behaviour modifiers.[34] In 1991 it was widely reported that researchers working for the British company Bodywise had discovered a scent that makes debt collection more efficient. It appears that persons who received bills treated with adrostenone, a substance found in men's sweat, were 17 per cent more likely to pay up than those who received odour-free bills. Bodywise reportedly patented the odorant and was already offering it to debt collecting agencies for some 3,000 pounds sterling a gram.[35]

In 1992, a Chicago researcher made the news by announcing that he had perfected a formula for a scent that induced casino gamblers to increase their betting on slot machines by as much as 45 per cent. The Chicago firm Inscentivation Inc. acquired exclusive rights to market the odorant and claimed to have entered into negotiations with casino owners from far and wide interested in purchasing it for use on their slot machines.[36]

Some business executives concerned with maximizing the productivity of their workers have turned to fragrance as a means of controlling behaviour. In one Japanese company, citrus scents are used to stimulate workers in the morning and after lunch when they are starting into their tasks. In the late morning and afternoon, when employees' minds might be drifting off their work, floral scents help boost their concentration. At midday and in the evening, when employees have started to lag, woodsy scents, such as cedar or cypress, help relieve their tiredness. Workers thus ride a crest of fragrance, carefully modulated to keep them performing at peak levels throughout the day.[37]

The use of fragrance to augment people's sense of well-being is known as 'aromatherapy' or 'aromachology'. Increasing popular acceptance of this field means that scents may now be marketed not only as lifestyle enhancers, but as therapeutic agents. Certain scent blends may be especially created to affect mood in a certain way. At present, for example, a perfume called 'Asleep' is marketed to help travellers sleep on planes, and one called 'Awake' to help them wake up and overcome jet lag. Such 'therapeutic' scents differ from standard perfumes in that they are created primarily to be smelled by the user, not to be worn so that others may smell them. However, in many cases, the two effects of internal well-being and external appeal may be sucessfully combined.

With aromatherapy, fragrances are no longer only aesthetic,

they are functional. This new functional character of scent makes it possible to present smell as an active ingredient in products, and not just a superficial attribute. Estée Lauder, for example, has already launched a 'Sensory Therapy' line, with such products as a mint-scented 'Peace of Mind On-the-Spot Gel'. Marketers of perfumes have a new sales pitch for their products as well. The ad for Bijan's *DNA* for men, for example, reminds consumers that 'science has recently proven that aromatic fragrances can significantly reduce anxiety'.

As the applications of scents multiply, fragrance merchants look forward to a boom in business. The president of the Fragrance Research Fund happily proclaims: 'It looks like an unlimited wealth of products is lurking just around the corner.'[38] It seems that the elimination of smell from cultural discourse – odours are not a topic of discussion in modernity – has been replaced by its elaboration through commercial discourse. Marketers have discovered what academics and other arbiters of culture have ignored: smell matters to people.

ARTIFICIAL FLAVOURS

'I think it's the best blueberry flavor that's ever been made. And there's not a scrap of blueberry in it', boasts the head flavorist at International Flavors & Fragrances Inc. in an article in the *Smithsonian*.[39] This remark provides a fitting introduction to the subject of artificial flavours[40] and their role in modern food.

Artificial flavours were invented in the late nineteenth century, but didn't become prevalent until the 1960s. At first such flavours were used to add taste and aroma to a limited range of foods – candies and beverages, for example – and to provide inexpensive substitutes for certain spices. No one expected them to completely replace natural flavours. Flavour engineers were accordingly modest about their achievements. A flavour company bulletin from the 1950s reads: 'We are proud to announce our new improved cherry flavor; of course it is still no match for Mother Nature's.'[41]

In the 1960s, however, flavorists set out to recreate virtually the whole spectrum of food flavours, from fruits and vegetables to meats. They have not been completely successful: some flavours, notably chocolate, coffee and bread, have eluded accurate simulation. None the less, the majority of the food on supermarket shelves today has at least some artificial flavouring. The only reason why synthetic versions of even more foods are not available is not because imitations of their flavours are lacking, but because their texture and appearance has thus far proved difficult to duplicate: 'If manufacturers manage to mold a chicken shape from vegetable protein, [flavorists] can dress it immediately with imitation chicken breast flavor, chicken fat flavor, chicken skin flavor and basic chicken flavor.'[42]

Major advances in flavour simulation have been made possible by the invention of the gas liquid chromatographer, which can chart the individual constituents of a particular flavour. This enables flavorists to identify the components of a flavour and reproduce them using chemicals known as 'flavomatics'. Not all the components or 'notes' present in the natural flavour are reproduced, only those deemed to be essential to its characteristic savour – perhaps twenty-five out of hundreds. In fact, by emphasizing certain flavour notes and downplaying or eliminating others, flavorists can create flavours that taste fuller and more palatable – at least to modern tastes – than the original.

The world of flavomatics has its unsavoury side as well, made up of an array of unappetizing flavour notes. 'Sweaty notes', for instance, are essential to the composition of such flavours as imitation rum and butterscotch, while 'fecal notes' give a full-bodied edge to cheese and nut flavours. Processed fruit flavours include a burnt undertone to mimic the effects of cooking. Artificial canned tomato flavours, in turn, must include the tinny taste consumers have come to expect from canned foods.[43] It is such graphic touches that enable artificial flavours to create a virtual reality of smell and taste.

Flavorists, indeed, are no longer willing to take second place to Mother Nature. Ads for their products now boast such wonders as coconut flavour that 'tastes more like coconuts than coconuts'.[44] In this enchanted garden of ideal synthetic savours, natural flavours intrude like a bad memory. Next to a well-made, sweet, full, artificial orange flavour, a real orange will likely taste sour and bland – a poor imitation. Furthermore, nature is unreliable and inconstant. Fruits and vegetables will only be in season during certain times of the year, and their quality will vary from crop to crop. Artificial fruit and vegetable flavours are available year-round and are always at their peak. Fruits and vegetables grow old and decay, they carry dirt, they often have unpalatable peels.

Their synthetic essences, in contrast, are pristine and unchanging: 'pure' flavour.[45]

Some artificial flavours have no counterpart in nature. The cola flavour found in soft drinks, for example, is the invention of an American pharmacist and tastes nothing like the cola nuts from which it takes its name.[46] Such purely synthetic creations, called 'fantasy flavours', are rare, however, as most artificial flavours are based on natural products.

Working against the complete triumph of artificial flavours is the public's growing distrust of synthetic food additives, which they fear may be carcinogenic or otherwise harmful. Such concerns notwithstanding, however, modern consumers have come to prefer strong and straightforward synthetic savours to their more subtle and complex natural counterparts. The trend is towards 'larger-than-life' flavours, especially popular among the young.[47]

At first, the flavour industry tried to counter concerns over the safety of its chemical creations by questioning the safety of consuming the products of Mother Nature. An industry report from the 1970s thus describes natural foods as, 'a wild mixture of substances created by plants or organisms for completely different non-food purposes – their survival and reproduction', which 'came to be consumed by humans at their own risk'.[48] Artificial flavours, on the other hand, the argument ran, have been tailor-made for human consumption.

Much as it desired to, however, the flavour industry has been unable to stem the 'back to nature' tide. In recent years researchers have accordingly been working to find alternatives to artificial flavours which will be able to satisfy both the public's demand for strong flavours and its desire for natural foods. One project involves engineering super-flavourful fruits and vegetables especially for use in the flavour industry.[49]

Of course, there is no real risk of artificial flavours definitively losing favour in the West. For one thing, the health concerns which prompt consumers to fear food additives also make them avid for safe substitutes for such suspect natural substances as sugar, fat and salt. For another, while artificial flavours can be used to give appealing savours to foods which are low in nutritive value (but big in profits) – such as soft drinks – they can also do the same to highly nourishing, low-cost foods – such as soybean products – making artificial flavours a potentially valuable tool in the struggle for a more equitable food order. Perhaps most importantly, modern consumers have become too accustomed to having a wide range of flavours at their disposal to willingly give up a number of their favourites, simply because their availability in a natural form is limited and their artificial counterparts suspect.[50]

Many flavours (such as vanilla and maple) are currently known to the general public only in their synthetic forms. This may increasingly be the case in the future, as artificially flavoured foods become more common worldwide. To quote a flavorist:

> In 20 years ... I'll bet you that only 5 percent of the people will have tasted fresh strawberry, so whether we like it or not, we people in the flavor industry will really be defining what the next generation thinks is strawberry. And the same goes for a lot of other foods that will soon be out of the average consumer's reach.[51]

For all those born into this new world of designer flavours, the scents and savours of dinner will often likely originate not in nature, but in laboratory vials, numbered and stored in an industrial flavour bank.

TRADEMARK SCENTS

When a particular scent is created for and becomes identified with a product, it is in the interest of the producer to secure legal control over that scent, so that imitation products cannot be marketed with the same smell. As we have seen, even when odour is simply an added attribute to a product, consumers often consider it to be indicative of the product's essential worth. By presenting itself with the same scent as a successful product – say a shampoo – a copycat product could thereby convince consumers that it was of equal quality and effectiveness.

While names and visual designs can readily be trademarked, however, odours pose more of a problem. How can one register an odour? How can there be property in 'air'? In order to avoid this difficulty, perfumers and flavorists have traditionally been very secretive regarding the composition of their olfactory creations. If no one knows the fragrance formula, the argument went, no one will be able to sucessfully recreate it:

A [fragrance] formula is like a bond payable to the bearer. There is no practical patent or copyright protection for fragrance formulations. Thus the only way in which the fragrance supplier can make sure that he and not someone else will collect the returns on his investments is by safely locking away his formulas, literally and figuratively.[52]

But modern technology, such as gas chromatography, has now made it possible to crack closely guarded fragrance formulas through chemical analysis. Very good imitation scents can therefore be produced by any company with the expertise and the necessary instruments. An expensive perfume, for example, which has required a large investment on the part of its manufacturer to create and market, can be closely copied by a competitor and sold at a fraction of the price. This practice can be observed in the current flood of imitation prestige perfumes on the market, their labels advertising to the consumer that they smell 'like Obsession', or 'like Giorgio', and so on.

This usurpation of profitable olfactory identities by hard-nosed competitors has intensified concern among fragrance manufacturers over how to establish legal control over their scents. The question of whether there can be property in odours was recently brought before the courts in an American case that involved a manufacturer of yarns and threads seeking to register the scent of Plumeria blossoms as a trademark for a product line called 'Clarke's Distinctive Sof-Scented Yarns'. The successful outcome of this case has significantly enhanced the commercial value of smells. The case, however, also reveals some of the difficulty the law of trademarks has in comprehending the signifying power of smells.[53]

A trademark is a device or symbol designed to serve as an indication of the origin of certain goods or services. According to conventional legal doctrine, trademarks need to be protected so as to safeguard consumers from deception and confusion, for example by low-cost imitators. The grant to a particular company of an exclusive right to the use of a particular identifying symbol, however, must not interfere with the ability of rival companies to compete. Competitive advantages may accrue from a mark or device that enhances the utility of the product, has a certain aesthetic appeal, or otherwise constitutes a characteristic which, if protected, would severely limit the number of options available to rivals to identify their goods. The ideal trademark is one, such as the name 'Exxon', that has no informational, aesthetic, or sensory value to consumers – except that of source identification.

Of course, marketers are interested in bending these rules as far as possible and thereby gaining every possible competitive advantage for their products. Concern has therefore been expressed over the possible consequences and implications of allowing odours to be licensed as trademarks. One legal commentator, for example, has raised the spectre of a manufacturer of leather shoes being allowed to register the smell of leather.[54] The difficulty in this case is that trademarking the natural odour of a product would make it more costly for other manufacturers to compete, for they would be compelled to go to the additional expense of obscuring the natural odour. Furthermore, in the case of manufacturers being obliged to change the natural scent of their product because it has already been registered by another company, buyers might then avoid the product simply because they are unable to recognize the material of which it is made from the smell.[55]

When a product fragrance is not natural to the product but an artificial attribute – such as a lilac scent added to detergent – the situation is less problematic, according to the legal experts. Competitors would be free to add their own floral, though not lilac, scents to their detergents, or to leave them without any added fragrance. The trademarking of such added fragrances, however, would draw attention to the fact that they *are* simply added attributes and not intrinsic to the product. This could in certain cases detract from the effectiveness of the scents as indicators of product worth.

Regardless of the legal niceties, product scents make up an increasingly large part of our modern olfactory experience and vocabulary. In one study, close to a thousand individuals of different ages, ranging from their twenties to their seventies, were approached in a Chicago shopping mall and asked the question: what odour causes you to become nostalgic? The responses, grouped by the decade in which the respondent was born, revealed a definite trend away from 'natural' odours and towards 'artificial' ones, or scents associated with commercial products. People born in the 1920s, 30s and 40s, said that such odours as rose, burning leaves, hot chocolate, cut grass and ocean air made them feel nostalgic. Persons born during the 1960s and 70s, in contrast, grow nostalgic at such scents as Downy fabric softener, hair spray, Play-Doh, suntan oil, Cocoa Puffs, and candy cigarettes.[56] Thus, the scent of hot chocolate fondly remembered by older generations is replaced by that of Cocoa Puffs among the younger, and ocean air by suntan oil.

In another study, reported in *Psychology Today*, researchers asked a group of over a hundred subjects to identify a number of odours. To the researchers' surprise, product scents such as baby powder, crayons and bubble gum proved to be more easily identifiable than such distinctive natural odours as coffee and lemon. Furthermore, subjects would almost invariably associate a brand name with a product scent: *Johnson's* baby powder, *Crayola* crayons, *Bazooka* bubble gum.[57] Just as our twentieth-century audiovisual imagination has been colonized by advertising logos, slogans and images, so it appears our modern olfactory consciousness will grow increasingly redolent with the odours of trademark scents – the aroma of commodities.

SMELL: THE POSTMODERN SENSE?

In our postmodern world smell is often a notable (or, increasingly, scarcely noticed) absence. Odours are suppressed in public places, there are no smells on television, the world of computers is odour-free, and so on. This olfactory 'silence' notwithstanding, smell would seem to share many of the traits commonly attributed to postmodernity. Let us make a comparison.

The postmodern era we live in is characterized by a loss of faith in universalist myths, such as Christianity or Progress, and a corresponding emphasis on the personal and local, on allegiance to one's own group. The breakdown of social structures, including language, encourages border-crossing (or simply lane-hopping) between such formerly rigid cultural categories as 'art' and 'life' or 'male' and 'female'. The past irrelevant, the future uncertain, postmodernity is a culture of 'now', a pastiche of styles and genres which exists in an eternal present. Postmodernity is also a culture of imitations and simulations, where copies predominate over originals and images over substance. The driving power of postmodernity is consumer capitalism, the endless production of goods and their investment with a quasi-religious aura of desirability.[58]

How does smell also exemplify these characteristics? First of all, odours are, by nature, personal and local. This enables olfactory values to be used to reinforce the tribal allegiances of post-modernity, in which the 'goodness' of one's own group is contrasted with the 'foulness' of others. At the same time, smells resist containment in discrete units, whether physical or linguistic; they cross borders, linking disparate categories and confusing boundary lines. Furthermore, smell, like taste, is a sensation of the moment, it cannot be preserved. We do not know what the past smelled like, and in the future our own odour will be lost. While odours cannot be preserved, however, they *can* be simulated. Commercially produced synthetic odours pervade the marketplace, enveloping consumer goods in ideal olfactory images.

This last point is best illustrated by an analysis of the industry of artificial flavours. The widespread replacement of natural flavours with artificial imitations which we find in the contemporary food industry exemplifies how, in Jean Baudrillard's words, the world has come to be 'completely catalogued and analyzed and then artificially revived as though real'.[59] Artificial flavours are created by the synthetic reproduction of individual flavour notes present in the original natural flavours. The flavorist may thus be regarded as the arch-agent in the process of production outlined by Baudrillard where: 'The real is produced from miniaturized units ... and with these it can be reproduced an indefinite number of times.'[60]

Ironically, in order to create 'larger-than-life' savours, flavorists actually reduce the number of components present in the natural flavour. By reproducing only those notes deemed essential to a flavour's characteristic taste and smell, they are able to produce a heightened sensation of that flavour. Artificial flavours are consequently at once much less than their originals and much more. Our contemporary craving for larger-than-life flavour is reminiscent of the medieval appetite for spices. While spices brought medievals a taste of Eden, however, artificial flavours are reminiscent rather of Disneyland, a synthetic paradise of consumer delights.[61]

The recession from and reinvention of reality occasioned by artificial flavours is aptly symbolized by the way *Coca-Cola* with its undeniably artificial flavour, is paraded as 'the real thing'. *Coke* is not a real thing in that it is not natural. *Coke* is an artificial thing. None the less, in that the reality of our world is increasingly defined and created for us by artifice (television, cinema, advertising), *Coca-Cola is* real. Put otherwise, having no basis in nature, *Coca-Cola* is able to represent the new, infinitely desirable, imagineered reality, with no tattered shreds of the shabby old reality clinging to it.

Smell, as we have seen in previous chapters, was considered an important force in the premodern West. This fact has not been lost on fragrance marketers, who make use of many of the ancient associations of smell – with magic, with sexual power, with healing – in order to promote their products. The very commercial process which ostensibly promotes traditional olfactory meanings, however, obviates the possibility of any real return to them, by transforming images of olfactory power into advertising copy.

In the past, essences were indicative of the intrinsic worth of the substances from which they emanated. Indeed, to encounter a scent was to encounter proof of a material presence, a trail of existence which could be traced to its source. Today's synthetic scents, however, are evocative of things which are not there, of presences which are absent: we have floral-scented perfumes which were never exhaled by a flower, fruit-flavoured drinks with not a drop of fruit juice in them, and so on. These artificial odours are a sign without a referent, smoke without fire, pure olfactory image.

This then is the manner in which smell, denied and ignored by scholars of modernity, can be called a 'postmodern' sense. Postmodernity, however, in no way allows for a full range of olfactory expression. Odours are rather eliminated from society and then reintroduced as packaged agents of fantasy, a means of recovering or recreating a body, an identity, a world, from which one has already been irrevocably alienated. The question is, will smell, seduced by an endless procession of olfactory simulacra, succumb to its postmodern fate, or will it – ever elusive – transcend its postmodern categorizations to remind us of our organic nature and even hint at a realm of the spirit.

Notes

Introduction

1 O. Sacks, *The Man Who Mistook His Wife for a Hat*, London, Duckworth, 1987, p. 159.

2 A. Synnott, 'Roses, Coffee, and Lovers: The Meanings of Smell', unpublished manuscript, 1993.

3 This last complaint was expressed in verse by E. B. White in his poem 'To a Perfumed Lady at a Concert':

> Madam reeking of the rose,
> Red of hair and pearl of earring,
> I came not to try my nose,
> I was there to try my hearing.
> Lost on me the whole darn concert.

E. B. White, *The Fox of Peapack and Other Poems*, New York, Harper & Brothers, 1938, p. 63.

4 See T. Engen, *Odor Sensation and Memory*, New York, Praeger, 1991, p. xii. On private codes of olfactory meaning see U. Almagor, 'Odors and Private Language: Observations on the Phenomenology of Scent', *Human Studies*, 1990, vol. 13, pp. 253–74.

5 See B. Schaal and R. Porter, ' "Microsmatic Humans" Revisited: The Generation and Perception of Chemical Signals', *Advances in the Study of Behavior*, 1991, vol. 20, pp. 135–9; R. Doty, 'Olfactory Communication in Humans', *Chemical Senses*, 1981, vol. 6, no. 4, pp. 351–76.

6 For instance, respondents to the Concordia smell survey ranked smell as the sense which was least important to them, making comments such as 'Makes no great difference whether I can smell things or not,' and 'I am not very dependent on this sense, and I don't smell things very well.' After having their olfactory consciousness raised by the survey some respondents changed their minds about the importance of smell. One wrote, for example: 'Ironically, after completing this questionnaire I realize how important smell is to one's life.'

7 See E. C. Carterette and M. P. Friedman (eds), *Handbook of Perception*, vol. 6A: *Tasting and Smelling*, New York, Academic Press, 1978; S. van Toller and G. Dodd (eds), *Perfumery: The Psychology and Biology of Fragrance*, London, Chapman & Hall, 1988.

8 The way in which olfaction is treated as a biological and psychological phenomenon in the modern West is itself a cultural phenomenon.

9 For a recent attempt to rationalize the 'decline of smell' among humans in evolutionary terms see D. M. Stoddart, *The Scented Ape*, Cambridge, Cambridge University Press, 1990.

10 P. Süskind, *Perfume: The Story of a Murderer*, J. Woods (trans.), New York, Alfred A. Knopf, 1986.

11 See D. Howes and M. Lalonde, 'The History of Sensibilities: Of the Standard of Taste in Mid-Eighteenth Century England and the Circulation of Smells in Post-Revolutionary France', *Dialectical Anthropology*, 1992, vol. 16, pp. 125–35.

12 P. Fauré, *Parfums et aromates de l'antiquité*, Paris, Fayard, 1987, p. 13.

13 Martial, *Epigrams*, W. Kerr (trans.), Cambridge, Mass., Harvard University Press, 1961, vol. 2, bk. 11: 8, p. 245.

14 A. Corbin, *The Foul and the Fragrant: Odor and the French Social Imagination*, M. Kochan, R. Porter and C. Prendergast (trans.), Cambridge, Mass., Harvard University Press, 1986.

15 G. Orwell, *The Road to Wigan Pier*, London, Victor Gollancz, 1937, pp. 159–60; emphasis in original.

1 The aromas of antiquity

1 Pliny, *Natural History*, H. Rackham (trans.), London, William Heinemann, 1960, vol. 4, bk. 13, p. 99. As this chapter is intended for a general readership, rather than classicists, all references to classical sources are from English translations.

2 On this see C. Classen, *Worlds of Sense: Exploring the Senses in History and Across Cultures*, London, Routledge, 1993, pp. 15–36.

3 Athenaeus, *The Deipnosophists or Banquet of the Learned*, C. D. Yonge (trans.), London, Henry G. Bohn, 1854, vol. 3, bk. 15, p. 1090.

4 Ibid., bk. 12, pp. 886–7.

5 Pliny, *Natural History*, vol. 6, bk. 21, p. 187.

6 Ibid., vol. 3, bk. 11, p. 175.

7 Ibid., vol. 5, bk. 17, pp. 27–9.

8 J. I. Miller, *The Spice Trade of the Roman Empire*, Oxford, Clarendon Press, 1969; N. Groom, *Frankincense and Myrrh: A Study of the Arabian Incense Trade*, London, Longman, 1981.

9 See P. Fauré, *Parfums et aromates de l'antiquité*, Paris, Fayard, 1987.

10 J. G. Griffiths (ed.), *Plutarch's 'De Iside et Osiride'*, Cardiff, University of Wales Press, 1970, ch. 80, p. 247.

11 Pliny, *Natural History*, vol. 4, bk. 13, pp. 101–9.

12 Ibid., p. 101.

13 Athenaeus, *Deipnosophists*, vol. 3, bk. 15, p. 1102.

14 Theophrastus, *Enquiry into Plants and Minor Works on Odours and Weather Signs*, A. Hort (trans.), London, William Heinemann, 1961, vol. 2, pp. 367–75.
15 Athenaeus, *Deipnosophists*, vol. 3, bk. 13, p. 970.
16 Ibid., bk. 15, p. 1101.
17 D. Grene and R. Lattimore (eds), *Sophocles*, Chicago, University of Chicago Press, 1954, p. 11.
18 See J. Corcopino, *Daily Life in Ancient Rome*, E. Lorimer (trans.), New Haven, Conn., Yale University Press, 1940, pp. 22–51.
19 G. Duckworth (ed.), *The Complete Roman Drama*, New York, Random House, 1942, vol. 1, p. 1011.
20 Pliny, *Natural History*, vol. 10, bk. 36, p. 141.
21 Martial, *Epigrams*, W. Kerr (trans.), Cambridge, Mass., Harvard University Press, 1961, vol. 2, bk. 10: 13, p. 163.
22 Pliny, *Natural History*, vol. 4, bk. 13, pp. 109–11.
23 Athenaeus, *Deipnosophists*, vol. 3, bk. 12, p. 885.
24 Martial, *Epigrams*, vol. 2, bk. 12: 32, p. 343.
25 S. Lilja, *The Treatment of Odours in the Poetry of Antiquity*, Commentationes Humanarum Litterarum 49, Helsinki, Societas Scientiarum Fennica, 1972, pp. 47–54. Lilja's book is an invaluable source of information on odour in classical literature.
26 Ibid., pp. 95–6.
27 Martial, *Epigrams*, vol. 1, bk. 4: 4, p. 233.
28 Ibid., vol. 2, bk. 9: 62, p. 121.
29 Athenaeus, *Deipnosophists*, vol. 1, bk. 3, p. 169.
30 Ibid., bk. 5, p. 314.
31 See, for example, Ovid, *Metamorphoses*, F. Miller (trans.), London, William Heinemann, 1960, vol. 1, bk. 8, p. 453.
32 Athenaeus, *Deipnosophists*, vol. 3, bk. 15, p. 1105.
33 Petronius, 'Satyricon', in *Petronius*, M. Heseltine (trans.), London, William Heinemann, 1919, p. 111.
34 *Lives of the Later Caesars: The First Part of the Augustan History*, A. Birley (trans.), Harmondsworth, Penguin, 1976, p. 306.
35 Duckworth, *Roman Drama*, vol. 1, p. 631.
36 Martial, *Epigrams*, vol. 1, bk. 3: 12, p. 171.
37 Athenaeus, *Deipnosophists*, vol. 1, bk. 4, p. 222.
38 Ibid., vol. 2, bk. 7, p. 455.
39 Ibid., p. 456.
40 Juvenal, 'Satires', in *Juvenal and Persius*, G. Ramsay (trans.), London, Heinemann, 1957, sat. 5, p. 81.
41 Athenaeus, *Deipnosophists*, vol. 2, bk. 9, p. 640.
42 Ibid., vol. 3, bk. 14, p. 1057.
43 Pliny, *Natural History*, vol. 4, bk. 13, p. 113.
44 Athenaeus, *Deipnosophists*, vol. 1, bk. 1, p. 49.
45 Duckworth, *Roman Drama*, vol. 1, p. 359.
46 Juvenal, 'Satires', sat. 6, p. 107.
47 Cited in C. J. S. Thompson, *The Mystery and Lure of Perfumes*, London, John Lane The Bodley Head, 1927, p. 73.
48 Corcopino, *Ancient Rome*, p. 265.

49 Lilja, *Treatment of Odours*, p. 50. See also W. Slater (ed.), *Dining in a Classical Context*, Ann Arbor, University of Michigan Press, 1991.
50 Athenaeus, *Deipnosophists*, vol. 2, bk. 10, p. 663.
51 Ibid., vol. 1, bk. 1, p. 28.
52 Ibid., bk. 4, p. 259.
53 Ibid., bk. 3, p. 184.
54 Ibid., bk. 2, p. 68.
55 Ibid., p. 71.
56 Ibid., p. 75.
57 Ibid., bk. 4, p. 259.
58 Lucretius, *On the Nature of Things*, C. Bailey (trans.), Oxford, Clarendon Press, 1910, bk. 2, p. 79.
59 R. Auget, *Cruelty and Civilization: The Roman Games*, London, Allen & Unwin, 1972.
60 Athenaeus, *Deipnosophists*, vol. 1, bk. 5, p. 312.
61 Ibid., pp. 313–20.
62 Martial, *Epigrams*, vol. 2, bk. 11: 8, p. 245.
63 Ibid., vol. 1, bk. 3: 65, p. 205.
64 *The Greek Anthology*, W. Paton (trans.), London, William Heinemann, 1960, vol. 1, bk. 5: 90, p. 171.
65 The Song of Solomon provides a Middle Eastern example of this complex, as evidenced in the following lines: 'Your love is more fragrant than wine, fragrant is the scent of your annointing oils, and your name is like those oils being poured out' (1: 1–2).
66 Duckworth, *Roman Drama*, vol. 1, p. 602.
67 Ibid., p. 285.
68 Athenaeus, *Deipnosophists*, vol. 3, bk. 15, p. 1099.
69 Ibid., bk. 13, p. 923.
70 Plutarch, *Makers of Rome: Nine Lives*, I. Scott-Kilvert (trans.), Harmondsworth, Penguin, 1965, p. 293.
71 Athenaeus, *Deipnosophists*, vol. 1, bk. 3, p. 184.
72 *Greek Anthology*, vol. 1, 5: 147, p. 199.
73 Lucretius, *Nature of Things*, bk. 4, p. 182.
74 Duckworth, *Roman Drama*, vol. 1, p. 102.
75 Athenaeus, *Deipnosophists*, vol. 3, bk. 15, p. 1070.
76 Plato, 'Symposium', in *Plato*, W. Lamb (trans.), London, William Heinemann, 1925, vol. 5, p. 155.
77 Martial, *Epigrams*, vol. 1, bk. 6: 93, p. 417.
78 On this see further Lilja, *Treatment of Odours*, pp. 124–44.
79 Martial has a similar epigram in *Epigrams*, vol. 1, bk. 5: 4, p. 233.
80 W. Ross (ed.), *The Works of Aristotle*, vol. VII: *Problemata*, Oxford, Clarendon Press, 1927, pp. 908a–8b.
81 Martial, *Epigrams*, vol. 1, bk. 3: 17, p. 175.
82 *Greek Anthology*, vol. 4, bk. 11: 239, p. 185.
83 Martial, *Epigrams*, vol. 2, bk. 13: 18, p. 397.
84 Juvenal, 'Satires', sat. 6, p. 119.
85 Martial, *Epigrams*, vol. 1, bk. 1: 28, p. 47.
86 Ibid., bk. 4: 4, p. 233.

87 Ross, *Problemata*, p. 908a.
88 Martial, *Epigrams*, vol. 2, bk. 11: 22, p. 255.
89 W. Oates and E. O'Neill Jr. (eds), *The Complete Greek Drama*, New York, Random House, 1938, vol. 2, p. 458.
90 Martial, *Epigrams*, vol. 1, bk. 1: 87, p. 83.
91 Pliny, *Natural History*, vol. 6, bk. 23, p. 523.
92 Martial, *Epigrams*, vol. 1, bk. 5: 4, p. 297.
93 Oates and O'Neill, *Greek Drama*, vol. 2, p. 885.
94 Pliny, *Natural History*, vol. 8, bk. 28, p. 41.
95 Martial, *Epigrams*, vol. 1, bk. 2: 12, p. 117.
96 Duckworth, *Roman Drama*, vol. 1, p. 630.
97 Martial, *Epigrams*, vol. 1, bk. 3: 55, p. 197.
98 On class divisions in antiquity see P. DuBois, *Centaurs and Amazons: Women and the Pre-History of the Great Chain of Being*, Ann Arbor, University of Michigan Press, 1982; and G. de Ste Croix, *The Class Struggle in the Ancient World*, Ithaca, N.Y., Cornell University Press, 1981.
99 Athenaeus, *Deipnosophists*, vol. 2, bk. 10, p. 693.
100 Lilja, *Treatment of Odours*, pp. 128–9. It was also the custom to place coins in the mouths of the dead so that they could pay their fare across the river Styx. J. M. C. Toynbee, *Death and Burial in the Roman World*, Ithaca, N.Y., Cornell University Press, 1971, p. 44.
101 Suetonius, *The Twelve Caesars*, R. Graves (trans.), Harmondsworth, Penguin, 1976, p. 285.
102 Martial, *Epigrams*, vol. 2, bk. 12: 59, p. 361.
103 Oates and O'Neill, *Greek Drama*, vol. 2, p. 542.
104 Duckworth, *Roman Drama*, vol. 1, p. 623.
105 A. D. Lindsay, (ed.), *Socratic Discourses by Plato and Xenophon*, London, J. M. Dent, 1910, p. 165.
106 Athenaeus, *Deipnosophists*, vol. 3, bk. 15, p. 1093.
107 Ibid., vol. 1, p. 377.
108 Oates and O'Neill, *Greek Drama*, vol. 2, p. 835.
109 Ibid., p. 836.
110 Athenaeus, *Deipnosophists*, vol. 3, bk. 15, p. 1096.
111 Ibid., p. 1097.
112 Ibid., p. 1096.
113 Duckworth, *Roman Drama*, vol. 1, p. 107.
114 Martial, *Epigrams*, vol. 1, bk. 3: 93, p. 223.
115 Horace, *Works of Horace*, C. Smart (trans.), London, George Bell & Sons, 1885, ode 12, p. 126.
116 *Greek Anthology*, vol. 1, bk. 5: 13, p. 135.
117 Juvenal, 'Satires', sats. 6 and 11, pp. 93, 233.
118 Ibid., sat. 6, p. 93.
119 Plautus, 'Poenulus', in *Plautus*, vol. 4, P. Nixon (trans.), London, William Heinemann, 1959, p. 27.
120 Lilja, *Treatment of Odours*, p. 134.
121 Ibid., p. 122.
122 For an analysis of the stereotypes expressed in the legend of Cleopa-

tra see L. Hughes-Hallet, *Cleopatra: Histories, Dreams and Distortions*, New York, Harper & Row, 1990.
123 Lucan, *The Civil War*, J. Duff (trans.), London, William Heinemann, 1957, vol. 1, bk. 6, p. 343.
124 Lucretius, *Nature of Things*, bk. 4, p. 182.
125 Athenaeus, *Deipnosophists*, vol. 2, bk. 7, p. 434.
126 See, for example, Lucan, *Civil War*, vol. 1, bk. 7, p. 431.
127 Martial, *Epigrams*, vol. 1, bk. 4: 4, p. 233.
128 Oates and O'Neill, *Greek Drama*, vol. 2, p. 691.
129 Ibid.
130 Ibid., p. 436.
131 See, for example, K. Nielsen, *Incense in Ancient Israel*, Leiden, E. J. Brill, 1986, p. 13.
132 Lucan, *Civil War*, vol. 1, bk. 9, p. 573.
133 Pliny, *Natural History*, vol. 4, bk. 15, p. 381.
134 Athenaeus, *Deipnosophists*, vol. 2, bk. 10, p. 664.
135 Ibid., vol. 3, bk. 12, p. 860.
136 Plutarch, *The Age of Alexander: Nine Greek Lives*, I. Scott-Kilvert (trans.), Harmondsworth, Penguin, 1973, p. 255.
137 Pliny, *Natural History*, vol. 3, bk. 8, p. 113.
138 Suetonius, *Twelve Caesars*, p. 37.
139 Plutarch, *Fall of the Roman Republic*, R. Warner (trans.), Harmondsworth, Penguin, 1972, p. 261.
140 W. Smith, *Dictionary of Greek and Roman Antiquities*, London, Walton & Maberly, 1853, pp. 359–61, 1166–7.
141 Athenaeus, *Deipnosophists*, vol. 3, bk. 15, p. 1111.
142 Ibid., p. 1098.
143 Ibid.
144 Ibid., p. 1077.
145 G. Majno, *The Healing Hand: Man and Wound in the Ancient World*, Cambridge, Mass., Harvard University Press, 1975, pp. 215–17.
146 Homer, *The Iliad*, A. Murray (trans.), London, William Heinemann, 1924, vol. 1, bk. 2, p. 105. See also Sophocles' play 'Philoctetes', in Oates and O'Neill, *Greek Drama*, vol. 1, pp. 533–607.
147 *Greek Anthology*, vol. 4, bk. 11: 165, p. 151.
148 Pliny, *Natural History*, vol. 6, bks. 20–3.
149 Athenaeus, *Deipnosophists*, vol. 1, bk. 1, p. 136; Lilja, *Treatment of Odours*, p. 100.
150 Athenaeus, *Deipnosophists*, vol. 1, bk. 3, pp. 170–1.
151 Nielsen, *Incense in Ancient Israel*, p. 8.
152 Ibid.
153 Ibid., p. 9.
154 Ibid.
155 Smith, *Greek and Roman Antiquities*, pp. 555–62; Pliny, *Natural History*, vol. 4, bk. 12, p. 61; vol. 6, bk. 11, p. 167; L. Friedländer, *Roman Life and Manners Under the Early Empire*, J. Freese and L. Magnus (trans.), London, Routledge & Kegan Paul, 1908, vol. 2, pp. 211–12.
156 Martial, *Epigrams*, vol. 2, bk. 10: 97, p. 229.
157 On Roman funerary rites see Toynbee, *Death and Burial*.

158 Martial, *Epigrams*, vol. 2, bk. 11: 54, pp. 277–9.
159 Persius, 'Satires', in *Juvenal and Persius*, G. Ramsay (trans.), London, William Heinemann, 1957, sat. 6, pp. 395–6.
160 Lucan, *Civil War*, vol. 1, bk. 7, p. 431.
161 Lilja, *Treatment of Odours*, p. 37.
162 Ibid.; C. H. Kahn, *The Art and Thought of Heraclitus: An Edition of the Fragments with Translation and Commentary*, Cambridge, Cambridge University Press, 1979, pp. 256–9.
163 Lucretius, *Nature of Things*, bk. 3, p. 125.
164 Smith, *Greek and Roman Antiquities*, p. 558.
165 Martial, *Epigrams*, vol. 1, bk. 5: 37, p. 323.
166 Homer, *Iliad*, vol. 2, bk. 15, p. 117.
167 Homer, 'Hymn to Demeter', in *The Homeric Hymns*, D. Hines (trans.), New York, Atheneum, 1972, p. 11.
168 Homer, 'Hymn to Aphrodite', in *Homeric Hymns*, p. 48.
169 Virgil, *The Aeneid*, J. Mantinband (trans.), New York, Frederick Ungar, 1964, bk. 1, p. 14.
170 Athenaeus, *Deipnosophists*, vol. 3, bk. 15, p. 1090.
171 M. Detienne, *The Gardens of Adonis: Spices in Greek Mythology*, J. Lloyd (trans.), Atlantic Highlands, N.J., The Humanities Press, 1977, pp. 69–70.
172 Homer, 'Hymn to Demeter', p. 4.
173 Homer, *Iliad*, vol. 2, bk. 14, p. 79.
174 See, for example, Homer, 'Hymn to Hermes', in *Homeric Hymns*, p. 40.
175 Oates and O'Neill, *Greek Drama*, vol. 2, pp. 673, 676.
176 Duckworth, *Roman Drama*, vol. 1, p. 822.
177 Oates and O'Neill, *Greek Drama*, vol. 2, p. 742.
178 Detienne, *Gardens of Adonis*, p. 41.
179 Athenaeus, *Deipnosophists*, vol. 1, bk. 1, p. 5.
180 Ibid., vol. 3, bk. 15, p. 1077.
181 Oates and O'Neill, *Greek Drama*, vol. 1, p. 797.
182 Ovid, *Metamorphoses*, vol. 1, bk. 4, p. 205.
183 Lilja, *Treatment of Odours*, p. 28.
184 Homer, *Iliad*, vol. 2, bk. 19, p. 339.
185 Ovid, *Metamorphoses*, vol. 2, bk. 15, p. 343.
186 Ibid., vol. 1, bk. 4, p. 197.
187 Ibid., vol. 2, bk. 10, pp. 87–99.
188 R. D. Archer-Hind (ed. and trans.), *The Timaeus of Plato*, New York, Macmillan & Co., 1888, pp. 243–7; Lilja, *Treatment of Odours*, pp. 10–13.
189 Lucretius, *Nature of Things*, bk. 2, p. 79.
190 B. S. Eastwood, 'Galen on the Elements of Olfactory Sensation', *Rheinisches Museum für Philologie*, 1981, vol. 124, pp. 268–89.
191 On this see Detienne, *Gardens of Adonis*, pp. 9–15 and Lilja, *Treatment of Odours*, p. 166.
192 Plautus, 'Pseudolus', in Duckworth, *Roman Drama*, vol. 1, p. 819.
193 Martial, *Epigrams*, vol. 2, bk. 12: 88, p. 381.
194 Oates and O'Neill, *Greek Drama*, vol. 2, p. 550.

195 Lucretius, *Nature of Things*, bk. 3, pp. 116–17.
196 Athenaeus, *Deipnosophists*, vol. 3, bk. 15, p. 1096.
197 Ibid., p. 1097.
198 Ibid., bk. 13, p. 976.

2 Following the scent

1 E. Atchley, *A History of the Use of Incense in Divine Worship*, London, Longmans, Green & Co., 1909, p. 85.

2 Cited in S. Harvey, 'The Fragrance of Sanctity: Incense and Spirituality in the Early Byzantine East', Dumbarton Oaks Public Lecture, 11 March 1992, p. 12.

3 T. McLaughlin, *Coprophilia or A Peck of Dirt*, London, Cassell, 1971, p. 11.

4 M. Grant, *The Fall of the Roman Empire: A Reappraisal*, Randar, Penn., Annenberg School Press, 1976, p. 219.

5 Atchely, *Incense in Divine Worship*, pp. 77–125.

6 See J.-P. Albert, *Odeurs de sainteté: la mythologie chrétienne des aromates*, Paris, Editions de l'Ecole des Hautes Etudes en Sciences Sociales, 1990.

7 Harvey, 'Fragrance of Sanctity', p. 11.

8 E. Cobham Brewer, *A Dictionary of Miracles*, Detroit, Gale Research Company, 1966, pp. 510–12; P. Guérin, *Vies des saints*, Paris, Bloud & Barral, 1878, 7th edn, vol. 5, pp. 216–29.

9 Cobham Brewer, *Dictionary of Miracles*, pp. 510–12.

10 T. Arnold (ed.), *Select English Works of John Wyclif*, vol. 1, Oxford, Clarendon Press, 1869, pp. 107–8.

11 A. L. Rowse, (ed.), *The Annotated Shakespeare*, New York, Clarkson N. Potter, 1978, vol. 2, *King Richard II*, act 3, scene 2, p. 321.

12 A. Neame, *The Holy Maid of Kent: The Life of Elizabeth Barton*, London, Hodder & Stoughton, 1971, pp. 334–5.

13 'A Description of a City Shower', in H. Williams (ed.), *The Poems of Jonathan Swift*, Oxford, Clarendon Press, 1958, vol. 1, p. 139.

14 L. Wright, *Clean and Decent: The Fascinating History of the Bathroom and the Water Closet*, Toronto, University of Toronto Press, 1960, p. 51.

15 G. Parfitt (ed.), *The Complete Poems of Ben Jonson*, Harmondsworth, Penguin, 1975, p. 91.

16 Moléon, *Rapports généraux*, cited in A. Corbin, *The Foul and the Fragrant: Odor and the French Social Imagination*, M. Kochan, R. Porter and C. Prendergast (trans.), Cambridge, Mass., Harvard University Press, 1986, p. 115.

17 J. D. Campbell (ed.), *The Poetical Works of Samuel Taylor Coleridge*, London, Macmillan, 1924, p. 452. European travellers often complained of the malodour of foreign cities and peoples. See, for instance, P. Braunstein, 'Toward Intimacy: The Fourteenth and Fifteenth Centuries', in G. Duby (ed.), *A History of Private Life*,

II: Revelations of the Medieval World, A. Goldhammer (trans.), Cambridge, Mass., Belknap Press, 1988, pp. 614–15.

18 J. Hogg, *London As It Is*, London, John Macrone, 1837, p. 228.

19 L.-S. Mercier, *Tableau de Paris*, cited in Corbin, *The Foul and the Fragrant*, p. 54.

20 Hogg, *London As It Is*, p. 229.

21 J. Harrington, *A New Discourse of a Stale Subject called the Metamorphosis of Ajax*, E. Dunno (ed.), London, Routledge & Kegan Paul, 1962 [1596], p. 85.

22 Ibid., p. 94.

23 Cited in W. Bell, *The Great Plague in London in 1665*, London, John Lane The Bodley Head, 1924, p. 341.

24 T. Norton, *Ordinal of Alchemy*, J. Reidy (ed.), Oxford, Oxford University Press, 1975, p. 64.

25 Ibid., p. 20.

26 A. Campbell, *The Black Death and Men of Learning*, New York, AMC Press, 1966, pp. 40, 44–5.

27 J. Papon, *Epoques mémorables de ce fléau*, 1800, cited in McLaughlin, *Coprophilia*, p. 16.

28 F. P. Wilson, *The Plague in Shakespeare's London*, Oxford, Oxford University Press, 1927, p. 4.

29 Ibid., p. 8; McLaughlin, *Coprophilia*, p. 25.

30 Wilson, *Plague in London*, p. 10.

31 E. R. Harvey, *The Inward Wits: Psychological Theory in the Middle Ages and the Renaissance*, London, The Warburg Institute, 1975, p. 27; R. Palmer, 'In Bad Odour: Smell and its Significance in Medicine from Antiquity to the Seventeenth Century', in R. Porter and W. F. Bynum (eds), *Medicine and the Five Senses*, Cambridge, Cambridge University Press, 1993, pp. 61–8.

32 F. P. Wilson (ed.), *The Plague Pamphlets of Thomas Dekker*, Oxford, Clarendon Press, 1925, pp. 34–5.

33 Cited in Wilson, *Plague in London*, p. 150.

34 D. Defoe, *A Journal of the Plague Year*, London, J. M. Dent, 1928 [1722], p. 102.

35 R. Reynolds, *Cleanliness and Godliness*, London, George Allen & Unwin, 1943, p. 75.

36 J. Swift, *Gulliver's Travels*, R. Greenberg (ed.), New York, W. W. Norton, 1961 [1726], p. 228.

37 Defoe, *Journal of the Plague Year*, p. 239.

38 Cited in McLaughlin, *Coprophilia*, p. 48.

39 L. Leminus, *The Touchstone of Complexions*, 1581, cited in A. Amherst, *A History of Gardening in England*, Detroit, Singing Tree Press, 1969 [1896], p. 164.

40 W. Harrison, *Description of England*, cited in E. Burton, *The Elizabethans at Home*, London, Secker & Warburg, 1958, p. 67.

41 Cited in Braunstein, 'Toward Intimacy', p. 539.

42 Cited in Wright, *Clean and Decent*, p. 49.

43 The Brunswick Court Regulations of 1589 stipulate: 'Let no one, whoever he may be, before, at or after meals, early or late, foul the

staircases, corridors, or closets with urine or other filth, but go to suitable, prescribed places for such relief.' Cited in N. Elias, *The Civilizing Process*, vol. 1: *The History of Manners*, E. Jephcott (trans.), New York, Pantheon, 1982, p. 131.

44 Corbin, *The Foul and the Fragrant*, p. 27.

45 R. Genders, *A History of Scent*, London, Hamish Hamilton, 1972, pp. 138–46.

46 *Apius and Virginia*, London, Chiswick Press, 1911 [1575], lines 280–3.

47 Ibid.; G. Vigarello, *Concepts of Cleanliness: Changing Attitudes in France since the Middle Ages*, J. Birrell (trans.), Cambridge, Cambridge University Press, 1988, p. 85.

48 Cited in Genders, *History of Scent*, p. 141.

49 C. J. S. Thompson, *The Mystery and Lure of Perfume*, London, John Lane The Bodley Head, 1927, p. 162.

50 Genders, *History of Scent*, pp. 139–46.

51 M. Kiernan (ed.), *Sir Francis Bacon: The Essayes or Counsels, Civill and Morall*, Oxford, Clarendon Press, 1985 [1625], pp. 140–1.

52 W. Cavendish, cited in Amherst, *History of Gardening*, p. 83.

53 For a fuller exposition of this topic see C. Classen, *Worlds of Sense: Exploring the Senses in History and Across Cultures*, London, Routledge, 1993, pp. 15–36.

54 G. Chaucer, *The Canterbury Tales*, D. Wright (trans.), Oxford, Oxford University Press, 1985, p. 171.

55 W. Schivelbusch, *Tastes of Paradise: A Social History of Spices, Stimulants, and Intoxicants*, D. Jacobson (trans.), New York, Pantheon Books, 1992, pp. 3–7; P. Pullar, *Consuming Passions: A History of English Food and Appetite*, London, Hamish Hamilton, 1970, pp. 97–8.

56 T. Austin (ed.), *Two Fifteenth-Century Cookery Books*, Oxford, Oxford University Press, 1964.

57 Schivelbusch, *Tastes of Paradise*, pp. 96–110.

58 L. Friedländer, *Roman Life and Manners Under the Early Empire*, J. Freese and L. Magnus (trans.), London, Routledge & Kegan Paul, 1908, p. 158. In this work, Friedländer compares the high-life of ancient Rome to that of Renaissance Europe.

59 Ibid., pp. 157–8.

60 Ibid., pp. 15–96; Pullar, *Consuming Passions*, p. 126.

61 Cited in Schivelbusch, *Tastes of Paradise*, p. 23.

62 Schivelbusch, *Tastes of Paradise*, pp. 22–3, 34–5

63 Pullar, *Consuming Passions*, pp. 95, 191.

64 Cited in Pullar, *Consuming Passions*, p. 151.

65 J. Swift, *Directions to Servants and Miscellaneous Pieces*, H. Davis (ed.), Oxford, Basil Blackwell, 1959 [1745].

66 Reynolds, *Cleanliness and Godliness*, p. 72.

67 Cited in McLaughlin, *Coprophilia*, p. 39.

68 Cited in F. Muir, *An Irreverent and Almost Complete Social History of the Bathroom*, New York, Stein & Day, 1983, p. 35.

69 Cited in Vigarello, *Concepts of Cleanliness*, p. 12.

70 Wright, *Clean and Decent*, p. 75.

71 Rowse, *Annotated Shakespeare*, vol. 1, *The Taming of the Shrew*, induction 1, scene 1, p. 124.

72 H. de Monteux, *Conservation de santé et prolongation de la vie*, Paris, 1572, p. 265, cited in Vigarello, *Concepts of Cleanliness*, p. 17.

73 Rowse, *Annotated Shakespeare*, vol. 1, *Much Ado About Nothing*, act 3, scene 2, p. 418.

74 L. Thorndike, *A History of Magic and Experimental Science*, vol. 7, New York, Colombia University Press, 1958, pp. 235, 264.

75 J. Nichols, *The Progresses of Queen Elizabeth*, London, John Nichols & Son, 1823, p. 319.

76 Thompson, *Mystery and Lure of Perfume*, pp. 143–6, 164–6; Corbin, *The Foul and the Fragrant*, pp. 74–7, 105, 196.

77 Wilson, *Plague Pamphlets*, p 29.

78 Corbin, *The Foul and the Fragrant*, pp. 66–75.

79 J. T. Shawcross (ed.), *The Complete Poetry of John Donne*, New York, New York University Press, 1968, p. 50.

80 From 'Amoretti and Epithalamion', in E. Spenser, *The Yale Edition of the Shorter Poems of Edmund Spenser*, W. Oram, E. Bjorvand, R. Bond, T. Cain, A. Dunlop, and R. Schell (eds), New Haven, Yale University Press, 1989, pp. 638–9.

81 L. C. Martin, (ed.), *The Poetical Works of Robert Herrick*, Oxford, Clarendon Press, 1956, p. 244.

82 Ibid., p. 210.

83 Williams, *Poems of Jonathan Swift*, vol. 2, p. 527.

84 Ibid., p. 584.

85 G. Della Casa, *Galateo*, cited in Elias, *Civilizing Process*, p. 131.

86 W. G. Ingram and T. Redpath (eds), *Shakespeare's Sonnets*, London, University of London Press, 1964, p. 125.

87 F. E. Hutchinson (ed.), *The Works of George Herbert*, Oxford, Clarendon Press, 1941, p. 94.

88 Ingram and Redpath, *Shakespeare's Sonnets*, p. 161.

89 Hutchinson, *Works of George Herbert*, p. 174.

90 Ibid., p. 181.

91 Hogg, *London As It Is*, p. 220.

92 Ibid., p. 225.

93 H. Gavin, *Sanitary Ramblings*, London, John Churchill, 1848, pp. 9, 20.

94 Ibid., p. 27.

95 V. Hugo, *Les Misérables*, C. Wilbour (trans.), New York, Modern Library, n.d., p. 1056.

96 Ibid., p. 1054.

97 Gavin, *Sanitary Ramblings*, p. 69.

98 Reynolds, *Cleanliness and Godliness*, p. 91.

99 Corbin, *The Foul and the Fragrant*, p. 213.

100 Ibid., pp. 79–104; McLaughlin, *Coprophilia*, p. 151.

101 Corbin, *The Foul and the Fragrant*, p. 22.

102 McLaughlin, *Coprophilia*, p. 148.

103 See, for example, Corbin, *The Foul and the Fragrant*, pp. 227–8.

104 G. Daignan, *Ordre de service des hôpitaux militaires*, Paris, 1785, p. 173, cited in Vigarello, *Concepts of Cleanliness*, p. 150.
105 Vigarello, *Concepts of Cleanliness*, p. 171.
106 Cited in McLaughlin, *Coprophilia*, p. 136.
107 C. H. Piesse (ed.), *Piesse's Art of Perfumery*, London, Piesse & Ludin, 1891, p. 32.
108 Vigarello, *Concepts of Cleanliness*, pp. 194–201.
109 Cited in Muir, *Social History of the Bathroom*, p. 25.
110 Vigarello, *Concepts of Cleanliness*, pp. 137–41.
111 Cited in P. Stalybrass and A. White, *The Politics and Poetics of Transgression*, Ithaca, N.Y., Cornell University Press, 1986, p. 139.
112 Thompson, *Mystery and Lure of Perfume*, p. 166.
113 See Classen, *Worlds of Sense*, pp. 87–93.
114 Ibid., p. 31.
115 Ibid., pp. 34–5.
116 Various works have dealt with this subject, including P. Poupon, *Mes dégustations littéraires: L'odorat et le goût chez les écrivains*, Dijon-Quetigny, Imprimerie Darantière, 1979; and H. J. Rindisbacher, *The Smell of Books: A Cultural-Historical Study of Olfactory Perception in Literature*, Ann Arbor, University of Michigan Press, 1992.
117 H. de Balzac, *Père Goriot*, A. J. Krailsheimer (trans.), Oxford, Oxford University Press, 1991 [1834], p. 5.
118 E. Zola, *Nana*, New York, Collier, 1962 [1880], p. 129.
119 C. Baudelaire, 'Lethe', D. Bell (trans.), in *The Flowers of Evil*, M. and J. Matthews (eds), New York, New Directions, 1989 [1857], p. 42.
120 O. Wilde, *The Picture of Dorian Gray*, New York, Modern Library, 1985 [1891], p. 148.
121 J.-K. Huysmans, *Against Nature*, R. Baldwick (trans.), London, Penguin, 1959 [1884], p. 120.
122 Ibid.
123 Baudelaire, 'Correspondences', R. Wilbur (trans.), in *Flowers of Evil*, p. 12.
124 Ibid.
125 Huysmans, *Against Nature*, p. 123.
126 Classen, *Worlds of Sense*, pp. 30–1.
127 M. Proust, *Remembrance of Things Past*, vol. I: *Swann's Way*, C. K. Moncrieff and T. Kilmartin (trans.), New York, Random House, 1981 [1913], p. 53.
128 Ibid., p. 51.
129 Corbin, *The Foul and the Fragrant*, pp. 35–48; H. Ellis, *Studies in the Psychology of Sex*, vol. 1, New York, Random House, 1942, [1899], part 3, pp. 48–81.
130 Corbin, *The Foul and the Fragrant*, pp. 1–61.
131 Ibid., pp. 67–70, 223–4.
132 E. B. de Condillac, *Treatise on the Sensations*, G. Carr (trans.), Los Angeles, University of Southern California Press, 1930, p. xxxi.
133 I. Kant, *Anthropology from a Pragmatic Point of View*, V. L. Dowdell (trans.), Carbondale and Edwardsville, Southern Illinois University

Press, 1978 [1798], 22, p. 46. Annick Le Guérer discusses attitudes towards olfaction in Western philosophy in *Scent: The Mysterious and Essential Powers of Smell*, R. Miller (trans.), New York, Turtle Bay Books, 1992, pp. 141–203.

134 C. Darwin, *The Descent of Man and Selection in Relation to Sex*, New York, D. Appleton, 1898, pp. 17–18.

135 S. Freud, *Civilization and its Discontents*, J. Strachey (trans.), New York, W. W. Norton, 1961, p. 46.

136 Ellis, *Psychology of Sex*, vol. 1, p. 62.

137 Ibid., p. 72.

138 Ibid., p. 107.

139 Ibid., pp. 72–3.

140 M. Nordau, *Degeneration*, New York, D. Appleton, 1902, p. 502.

141 Ibid.

142 Ibid., pp. 502–3.

143 A. Galopin, *Le parfum de la femme*, 1886, cited in Ellis, *Psychology of Sex*, p. 78.

144 In *The Descent of Man*, Darwin stated that 'the sense of smell is of extremely slight service, if any, even to the dark coloured races of men, in whom it is more highly developed than in the white and civilized races' (pp. 17–18).

145 Notably Claude Lévi-Strauss. See, for example, his book *The Raw and the Cooked: Introduction to a Science of Mythology*, vol. 1, J. and D. Weightman (trans.), New York, Harper & Row, 1969.

3 Universes of odour

1 The word osmology literally means 'theory of smell'. We use it here to refer to how societies order the cosmos through and in terms of concepts derived from olfaction.

2 A. R. Radcliffe-Brown, *The Andaman Islanders*, New York, The Free Press, 1964, pp. 119, 311–12. Scent calendars similar to the Andaman one have been reported for other South Sea island societies, such as Normanby and New Britain. See G. Roheim, *Psychoanalysis and Anthropology*, New York, International Universities Press, 1950, pp. 151–3; M. Panoff, 'The Notion of Time Among the Maenge People of New Britain', *Ethnology*, 1969, vol. 8, pp. 153–66.

3 Radcliffe-Brown, *Andaman Islanders*, p. 312.

4 Ibid., p. 119.

5 U. Almagor, 'The Cycle and Stagnation of Smells: Pastoralists–Fishermen Relationships in an East African Society', *RES*, 1987, vol. 14, pp. 106–21. The olfactory contrast between dry and wet seasons can indeed be very striking, as Marja-Liisa Swantz, who has done anthropological fieldwork in Tanzania, relates:

> I have experienced a lack of strong scents in the dry tropical bushland, but when the refreshing rains come, the world is satu-

rated with odour. Otherwise the smell associations are centred around the village, where smoke and food fumes fill the yard.

M.-L. Swantz, *Ritual and Symbol in Transitional Zaramo Society with Special Reference to Women*, Uppsala, Almquist & Wiksells, 1970, p. 235.

6 Radcliffe-Brown, *Andaman Islanders*, p. 311.
7 The notion of smellscapes is derived from the work of D. Porteous, particularly his *Landscapes of the Mind: Worlds of Sense and Metaphor*, Toronto, University of Toronto Press, 1990.
8 V. Pandya, 'Above the Forest: A Study of Andamanese Ethnoamenology, Cosmology and the Power of Ritual', Ph.D. thesis, University of Chicago, 1987; V. Pandya, 'Movement and Space: Andamanese Cartography', *American Ethnologist*, 1991, vol. 17, no. 4, pp. 775–97.
9 Pandya, 'Above the Forest'.
10 G. Lewis, *Knowledge of Illness in a Sepik Society: A Study of the Gnau, New Guinea*, London, Athlone Press, 1975, p. 46.
11 A. Gell, 'Magic, Perfume, Dream . . .' in I. M. Lewis (ed.), *Symbols and Sentiments*, London, Academic Press, 1977, p. 32.
12 G. Reichel-Dolmatoff, 'Tapir Avoidance in the Colombian Northwest Amazon', in G. Urton (ed.), *Animal Myths and Metaphors in South America*, Salt Lake City, University of Utah Press, 1985, pp. 124–5.
13 G. Reichel-Dolmatoff, 'Desana Animal Categories, Food Restrictions, and the Concept of Color Energies', *Journal of Latin American Lore*, 1978, vol. 4, no. 2, pp. 243–91.
14 Reichel-Dolmatoff, 'Tapir Avoidance', pp. 124–5.
15 Reichel-Dolmatoff, 'Desana Animal Categories', p. 273.
16 Ibid., pp. 271–4; Reichel-Dolmatoff, 'Tapir Avoidance', pp. 124–9.
17 Reichel-Dolmatoff, 'Desana Animal Categories', p. 271.
18 A. Seeger, *Nature and Society in Central Brazil: The Suya Indians of Mato Grosso*, Cambridge, Mass., Harvard University Press, 1981, pp. 93–120.
19 Ibid., pp. 107–15; A. Seeger, 'Anthropology and Odor: From Manhattan to Mato Grosso', *Perfumer & Flavorist*, 1988, vol. 13, pp. 41–8.
20 J. C. Crocker, *Vital Souls: Bororo Cosmology, Natural Symbolism, and Shamanism*, Tucson, University of Arizona Press, 1985, pp. 158–9.
21 Ibid., pp. 41–3.
22 Ibid.
23 M. Dupire, 'Des goûts et des odeurs: classifications et universaux', *L'Homme*, 1987, vol. 27, no. 4, pp. 11–14.
24 See D. Howes, 'Olfaction and Transition', in D. Howes (ed.), *The Varieties of Sensory Experience*, Toronto, University of Toronto Press, 1991, pp. 139–140.
25 Dupire, 'Des goûts et des odeurs', p. 8.
26 One wonders how much richer our understanding of some cultures would be if more anthropologists had, like the ones whose work we

have been relying on here, focused on eliciting the 'world-scent' (and 'world-sound', etc.) of the cultures they studied, instead of limiting themselves to describing the 'world-view'. On the importance of adjusting one's sensory *ratio* to conform to that of the culture under study see D. Howes, 'Sense and Non-Sense in Contemporary Ethno/Graphic Practice and Theory', *Culture*, 1991, vol. 11, nos. 1–2, pp. 65–76; C. Classen, *Worlds of Sense: Exploring the Senses in History and Across Cultures*, London, Routledge, 1993, pp. 104–5.

27 Crocker, *Vital Souls*, pp. 42, 56.

28 W. James, *The Listening Ebony: Moral Knowledge, Religion and Power Among the Uduk of Sudan*, Oxford, Clarendon Press, 1988, p. 72.

29 Reichel-Dolmatoff, 'Tapir Avoidance', p. 129.

30 Reichel-Dolmatoff, 'Desana Animal Categories', pp. 252, 278, 284.

31 G. Reichel-Dolmatoff, *Amazonian Cosmos: The Sexual and Religious Symbolism of the Tukano Indians*, Chicago, University of Chicago Press, 1971, p. 235.

32 Reichel-Dolmatoff, 'Desana Animal Categories', pp. 278–9.

33 K. Endicott, *Batek Negrito Religion: The World-View and Rituals of a Hunting and Gathering People of Peninsular Malaysia*, Oxford, Clarendon Press, 1979, pp. 74–6.

34 Ibid.

35 Ibid., pp. 62–3.

36 Ibid., p. 75.

37 Ibid., p. 63.

38 W. E. A. von Beek, 'The Dirty Smith: Smell as a Social Frontier among the Kapsiki/Higi of North Cameroon and North-Eastern Nigeria', *Africa*, 1992, vol. 62, no. 1, pp. 38–58. The Kapsiki are known as Higi in Nigeria.

39 Ibid.

40 Further in this vein, one wonders, for example, if Suya women concur with the way Suya men classify them as 'our rotten smelling property', or if they have not developed their own scale of olfactory values. Regarding this issue of internal cultural diversity, see further A. Corbin, 'Histoire et anthropologie sensorielle', *Anthropologie et Sociétés*, 1990, vol. 14, no. 2, pp. 13–24, 40.

41 See R. Harper, E. C. Bate Smith and D. G. Land, *Odour Description and Odour Classification*, London, J. & A. Churchill, 1968.

42 Dupire, 'Des goûts et des odeurs', pp. 8–14.

43 Ibid.

44 von Beek, 'The Dirty Smith', p. 43.

45 Ibid.

46 Reichel-Dolmatoff, 'Desana Animal Categories', pp. 271–2.

47 C. Classen, *Inca Cosmology and the Human Body*, Salt Lake City, University of Utah Press, 1993, p. 165.

48 Pandya, 'Above the Forest', p. 167.

49 Crocker, *Vital Souls*, pp. 64–5, 158–61.

50 Dupire, 'Des goûts et des odeurs', pp. 11–12; M. Dupire, 'Nomi-

nation, réincarnation et/ou ancêtre tutélaire? Un mode de survie: L'example des serer ndout (Sénégal)', *L'Homme*, 1982, vol. 22, no. 1, pp. 5–31.
51 Pandya, 'Above the Forest', pp. 114–5.
52 J. J. Meyer, *Sexual Life in Ancient India*, Delhi, Banarsidass, 1971, p. 183. Similar practices exist in Arabic cultures:

> Arabs consistently breathe on people when they talk. To the Arab good smells are pleasing and a way of being involved with each other. To smell one's friend is not only nice but desirable, for to deny him your breath is to act ashamed.

E. T. Hall, *The Hidden Dimension*, Garden City, N.Y., Anchor, 1969, pp. 159–60. Such practices differ significantly from what passes as acceptable behaviour in mainstream Western society, where to breathe on or smell someone is considered impolite. Hall suggests that the differing norms concerning personal distance which different societies have developed can be a source of friction in cross-cultural encounters, as persons alternately invade or withdraw from each others' olfactory space and by so doing communicate all sorts of messages of which they are not in the least conscious.

53 Reichel-Dolmatoff, 'Desana Animal Categories', p. 281; G. Reichel-Dolmatoff, *Basketry as Metaphor: Arts and Crafts of the Desana Indians of the Northwest Amazon*, Los Angeles, Occasional Papers of the Museum of Cultural History, University of California, 1985, pp. 24, 33.
54 Endicott, *Batek Negrito Religion*, pp. 77, 94.
55 M. Roseman, *Healing Sounds from the Malaysian Rainforest: Temiar Music and Medicine*, Berkeley, University of California Press, 1991, pp. 21, 37.
56 Seeger, *Nature and Society*, pp. 112–15.
57 Reichel-Dolmatoff, 'Desana Animal Categories', pp. 271–4, 279; Reichel-Dolmatoff, 'Tapir Avoidance', pp. 124–5.
58 Endicott, *Batek Negrito Religion*, pp. 38–41, 124–5.
59 Crocker, *Vital Souls*, pp. 37, 122.
60 Ibid., pp. 37, 85, 122.
61 Ibid., pp. 313–28.
62 Pandya, 'Above the Forest'.
63 Ibid., pp. 277–8.
64 Ibid., p. 278.
65 G. Calame-Griaule, *Words and the Dogon World*, D. Lapin (trans.), Philadelphia, Institute for the Study of Human Issues, 1986, pp. 32, 36–40, 132, 308, 311–13.
66 This account of the Chinese system of correspondences and the table are based on the discussion in J. Needham, *Science and Civilisation in China*, vol. 2, Cambridge, Cambridge University Press, 1969, pp. 232ff. See also M. Porkert, *The Theoretical Foundations of Chinese Medicine*, Cambridge, Mass., MIT Press, 1974.
67 Cited in J. B. Henderson, *The Development and Decline of Chinese Cosmology*, New York, Columbia University Press, 1984, p. 21.

68 Reichel-Dolmatoff, 'Desana Animal Categories', pp. 256–78.
69 Ibid.
70 Reichel-Dolmatoff, *Amazonian Cosmos*, p. 123.
71 Ibid., pp. 111–16, 142; G. Reichel-Dolmatoff, 'Brain and Mind in Desana Shamanism', *Journal of Latin American Lore*, 1981, vol. 7, no. 1, p. 91.
72 Ibid., pp. 82–6.
73 See D. Howes (ed.), *The Varieties of Sensory Experience: A Sourcebook in the Anthropology of the Senses*, Toronto, University of Toronto Press, 1991.

4 The rites of smell

1 D. Shulman, 'The Scent of Memory in Hindu South India', *RES*, 1987, vol. 13, p. 132.
2 U. Almagor, 'The Cycle and Stagnation of Smells: Pastoralists–Fishermen Relationships in an East African Society', *RES*, 1987, vol. 14, p. 109.
3 Shulman, 'Scent of Memory', p. 135.
4 W. H. I. Bleek and L. C. Lloyd (eds), *Specimens of Bushman Folklore*, London, George Allen, 1911, p. 193.
5 B. Malinowski, *The Sexual Life of Savages in North-Western Melanesia*, London, Harcourt, Brace & World, 1929, pp. 312, 378.
6 S. Petit-Skinner, *The Nauruans*, San Francisco, Macduff Press, 1981, p. 98.
7 Ibid., p. 9.
8 Malinowski, *Sexual Life of Savages*, p. 250.
9 G. Reichel-Dolmatoff, *Amazonian Cosmos: The Sexual and Religious Symbolism of the Tukano Indians*, Chicago, University of Chicago Press, 1971, p. 120.
10 A. Kanafani, *Aesthetics and Ritual in the United Arab Emirates: The Anthropology of Food and Personal Adornment among Arabian Women*, Beirut, American University of Beirut, 1983, pp. 42–3, 47–8.
11 Ibid., pp. 41, 44–5.
12 Ibid., pp. 41, 93.
13 Ibid., p. 90.
14 Ibid., pp. 43, 47, 48, 50.
15 Ibid., pp. 46, 48–50.
16 Ibid., pp. 46–7, 50.
17 Ibid., pp. 34–40, 81–9.
18 W. G. Palgrave, *Narrative of a Year's Journey Through Central and Eastern Arabia*, London, Macmillan, 1866, vol. 2, p. 26.
19 Ibid., pp. 23–5.
20 Ibid., pp. 26, 101.
21 Petit-Skinner, *Nauruans*, p. 97.
22 Kanafani, *Aesthetics and Ritual*, p. 42.
23 Ibid., p. 50.
24 Ibid., p. 47

25 E. A. Westermarck, *Ritual and Belief in Morocco, I*, London, Macmillan, 1926, pp. 234, 236, 280.
26 See E. Z. Vogt, *Tortillas for the Gods: A Symbolic Analysis of Zinacanteco Rituals*, Cambridge, Mass., Harvard University Press, 1976; P. Stoller and C. Olkes, *In Sorcery's Shadow: A Memoir of Apprenticeship Among the Songhay of Nigeria*, Chicago, University of Chicago Press, 1987, pp. 188–9; D. Howes, 'Olfaction and Transition', in *The Varieties of Sensory Experience: A Sourcebook in the Anthropology of the Senses*, D. Howes (ed.), Toronto, University of Toronto Press, 1991, p. 131.
27 S. Howell, *Society and Cosmos: Chewong of Peninsular Malaysia*, Oxford, Oxford University Press, 1984, pp. 24, 92.
28 Ibid., pp. 95–6.
29 K. Endicott, *Batek Negrito Religion: The World-View and Rituals of a Hunting and Gathering People of Peninsular Malaysia*, Oxford, Clarendon Press, 1979, pp. 143–4, 156–8.
30 S. Leacock and R. Leacock, *Spirits of the Deep: A Study of an Afro-Brazilian Cult*, Garden City, N.Y., Doubleday Natural History Press, 1972, p. 6.
31 Ibid., p. 147. For a similar example see M. Lambek, *Human Spirits: A Cultural Account of Trance in Mayotte*, Cambridge, Cambridge University Press, 1981, pp. 119–23.
32 J. Boddy, *Wombs and Alien Spirits: Women, Men and the Zar Cult in Northern Sudan*, Madison, University of Wisconsin Press, 1989, pp. 141, 153, 189.
33 Ibid., p. 248
34 Ibid., pp. 250, 269–309.
35 V. Pandya, 'Above the Forest: A Study of Andamanese Ethnoamenology, Cosmology and the Power of Ritual', Ph.D. thesis, University of Chicago, 1987, pp. 133, 136.
36 Ibid., pp. 139–40.
37 Ibid., pp. 168–78.
38 For a particularly rich discussion on this point see J.-L. Nakbi, 'Sémiologie des odeurs rituelles: Rites et rythmes olfactifs dans le sud tunisien', *Cahiers de Sociologie Economique et Culturelle*, 1985, vol. 4, pp. 117–36.
39 H. Forbes, *A Naturalist's Wanderings in the Eastern Archipelago*, London, Sampson, Low, Marston, Searle & Rivington, 1885, p. 315.
40 Boddy, *Wombs and Alien Spirits*, pp. 106–7.
41 J. C. Crocker, *Vital Souls: Bororo Cosmology, Natural Symbolism and Shamanism*, Tucson, University of Arizona Press, 1985, pp. 52–9.
42 A. Seeger, *Nature and Society in Central Brazil: The Suya Indians of Mato Grosso*, Cambridge, Mass., Harvard University Press, 1981, pp. 112–15.
43 Reichel-Dolmatoff, *Amazonian Cosmos*, pp. 29, 143; G. Reichel-Dolmatoff, 'Desana Animal Categories, Food Restrictions, and the Concept of Color Energies', *Journal of Latin American Lore*, 1978, vol. 4, no. 2, pp. 277–8.

44 A. Meigs, *Food, Sex and Pollution: A New Guinea Religion*, New Brunswick, N.J., Rutgers University Press, 1984, p. 32.
45 Ibid., pp. 70–1.
46 See further, T. Buckley and A. Gottlieb (eds), *Blood Magic: The Anthropology of Menstruation*, Berkeley, University of California Press, 1988.
47 Crocker, *Vital Souls*, p. 60.
48 Kanafani, *Aesthetics and Ritual*, pp. 50, 74–5.
49 Almagor, 'Cycle and Stagnation of Smells', pp. 111, 115.
50 Pandya, 'Above the Forest', p. 211.
51 Ibid., pp. 221–2, 226.
52 Ibid., pp. 221, 259, 271–87.
53 Ibid., pp. 283–8, 321.
54 Kanafani, *Aesthetics and Ritual*, p. 78.
55 Ibid., pp. 78–9.
56 Ibid., p. 79.
57 Boddy, *Wombs and Alien Spirits*, pp. 106–7, 329–30. For a comparable custom in Tunisia see Nakbi, 'Sémiologie des odeurs rituelles', pp. 117–36.
58 Boddy, *Wombs and Alien Spirits*, pp. 106–7, 330.
59 See Howes, 'Olfaction and Transition'.
60 M. Kahn, *Always Hungry, Never Greedy: Food and the Expression of Gender in a Melanesian Society*, Cambridge, Cambridge University Press, 1986, p. 36.
61 Ibid., pp. 101, 118.
62 B. Malinowski, *Coral Gardens and Their Magic*, Bloomington, Indiana University Press, 1965, vol. 1, p. 232.
63 Endicott, *Batek Negrito Religion*, pp. 12, 57–9.
64 Ibid., pp. 66, 136.
65 Reichel-Dolmatoff, *Amazonian Cosmos*, pp. 220–1.
66 Ibid.
67 Ibid., pp. 221–2; Reichel-Dolmatoff, 'Desana Animal Categories', pp. 274–5.
68 Reichel-Dolmatoff, *Amazonian Cosmos*, pp. 223–5; G. Reichel-Dolmatoff, 'Tapir Avoidance in the Colombian Northwest Amazon', in *Animal Myths and Metaphors*, G. Urton (ed.), Salt Lake City, University of Utah Press, 1985, pp. 120–1.
69 Pandya, 'Above the Forest', p. 16.
70 Ibid., pp. 148–9.
71 Ibid., p. 160.
72 Ibid., pp. 298–9.
73 Ibid., p. 91.
74 Ibid.
75 Kahn, *Always Hungry, Never Greedy*, p. 101.
76 Endicott, *Batek Negrito Religion*, p. 207.
77 Reichel-Dolmatoff, 'Tapir Avoidance', p. 123.
78 W. Wilbert, 'The Pneumatic Theory of Female Warao Herbalists', *Social Sciences and Medicine*, 1987, vol. 25, no. 10, p. 1141.
79 Ibid., pp. 1141–2.

80 Endicott, *Batek Negrito Religion*, pp. 106–8.
81 C. Classen and D. Howes, 'Aromatherapy in the Andes', *Dragoco Report*, 1993, vol. 40, no. 6, pp. 213–27.
82 Pandya, 'Above the Forest', p. 107.
83 Ibid.
84 Ibid., pp. 145–7.
85 M. Roseman, *Healing Sounds from the Malaysian Rainforest: Temiar Music and Medicine*, Berkeley, University of California Press, 1991, pp. 37–40.
86 A. Gebhart-Sayer, 'The Geometric Designs of the Shipibo-Conibo in Ritual Context', *Journal of Latin American Lore*, 1985, vol. 11, no. 12, pp. 143–74.
87 M.-L. Swantz, *Ritual and Symbol in Traditional Zaramo Society with Special Reference to Women*, Uppsala, Sweden, Almquist & Wiksells, 1970, pp. 234, 268, 307.
88 L. L. Wall, *Hausa Medicine*, Durham, N.C., Duke University Press, 1988, pp. 193, 315.
89 Gebhart-Sayer, 'Geometric Designs', pp. 161–2.
90 Endicott, *Batek Negrito Religion*, pp. 64, 114–20.
91 Ibid., pp. 49–50, 112–13.
92 Kanafani, *Aesthetics and Ritual*, pp. 79–80.
93 Crocker, *Vital Souls*, pp. 117, 160.
94 Pandya, 'Above the Forest', p. 88.
95 Ibid., p. 85.
96 Ibid., pp. 85–7, 157.
97 Ibid., pp. 156–9.
98 M. Leenhardt, *Do Kamo: Person and Myth in the Melanesian World*, B. Miller Gulati (trans.), Chicago, University of Chicago Press, 1979, pp. 48–9, 52.
99 Ibid., pp. 30–1.
100 Ibid., pp. 53–4.
101 S. Küchler, 'The Epidemiology of Imagery and the Epistemology of Smell in Northern New Ireland', paper presented at 87th Annual Meeting of the American Anthropological Association, 1988.
102 Ibid.
103 See further D. Howes, 'On the Odour of the Soul: Spatial Representation and Olfactory Classification in Eastern Indonesia and Western Melanesia', *Bijdragen tot de Taal-, Land- en Volkenkunde*, 1988, vol. 124, pp. 84–113.
104 Cf. J. T. Siegel, 'Images and Odors in Javanese Practices Surrounding Death', *Indonesia*, 1983, vol. 36, no. 1, pp. 1–15. In Java, the odour of the corpse must never be allowed to interfere with the visual image of the deceased.
105 A. Gell, 'Magic, Perfume, Dream . . .' in *Symbols and Sentiments*, I. M. Lewis (ed.), London, Academic Press, 1977, pp. 25–38.
106 Pandya, 'Above the Forest', p. 108.
107 F. B. Lamb, *Wizard of the Upper Amazon*, Boston, Houghton Mifflin, 1975, p. 90.
108 G. Reichel-Dolmatoff, *Beyond the Milky Way: Hallucinatory Ima-*

gery of the Tukano Indians, Los Angeles, UCLA Latin American Center Publications, 1978, pp. 11–12; Reichel-Dolmatoff, *Amazonian Cosmos*, p. 172.

109 Reichel-Dolmatoff, *Beyond the Milky Way*, p. 13.
110 Ibid., pp. 12–13; Reichel-Dolmatoff, *Amazonian Cosmos*, p. 46.
111 G. Reichel-Dolmatoff, 'Brain and Mind in Desana Shamanism', *Journal of Latin American Lore*, 1981, vol. 7, no. 1, pp. 94–5.
112 Reichel-Dolmatoff, *Amazonian Cosmos*, pp. 133–4.

5 Odour and power

1   A. H. Verrill, *Perfumes and Spices*, Boston, L. C. Page, 1940, p. 91.
2   C. Classen, *Worlds of Sense: Exploring the Senses in History and Across Cultures*, London, Routledge, 1993, pp. 87–90, 93.
3   Verrill, *Perfumes and Spices*, p. 97.
4   Ibid., p. 95.
5   Ibid., p. 195.
6   C. J. S. Thompson, *The Mystery and Lure of Perfume*, London, John Lane The Bodley Head, 1927, p. 151.
7   H. Williams (ed.), *The Poems of Jonathan Swift*, vol. 2, Oxford, Clarendon Press, 1937, p. 529.
8   Ibid., 'Strephon and Chloe', pp. 591–2.
9   A. Synnott, *The Body Social: Symbolism, Self and Society*, London, Routledge, 1993, pp. 73–102, 198–201; C. Ballerino Cohen, 'Olfactory Constitution of the Postmodern Body: Nature Challenged, Nature Adorned', in F. E. Mascia Lees and P. Sharpe (eds), *Tatoo, Torture, Mutilation and Adornment*, Binghampton, N.Y., State University of New York Press, 1992, pp. 48–78.
10 R. Doty, 'The Role of Olfaction in Man: Sense or Nonsense?', in S. H. Bartley (ed.), *Perception in Everyday Life*, New York, Harper & Row, 1972, pp. 149–50.
11 Cited in R. Bedichek, *The Sense of Smell*, Garden City, N.Y., Doubleday, 1960, p. 155.
12 See Synnott, *The Body Social*, pp. 194–202; and G. P. Largey and D. R. Watson, 'The Sociology of Odors', *The American Journal of Sociology*, 1977, vol. 77, no. 6, pp. 1021–34.
13 E. E. Cashmore, *The Logic of Racism*, London, Allen & Unwin, 1987, p. 86.
14 D. Hamilton, 'Chirac's remark on Immigrants Touches Raw Nerve in French Politics', *The Montreal Gazette*, Saturday, 22 June 1991, p. B8; M. Peyrot, 'La plainte du MRAP contre M. Chirac: Le procès des "odeurs" ', *Le Monde*, Friday, 31 January 1992, p. 7.
15 G. Orwell, *The Road to Wigan Pier*, London, Victor Gollancz, 1937, p. 159.
16 Ibid., p. 160.
17 W. S. Maugham, *On A Chinese Screen*, Oxford, Oxford University Press, 1927, pp. 153–4.
18 Orwell, *Wigan Pier*, p. 163.

19 Ibid., p. 160.
20 J. Dollard, *Caste and Class in a Southern Town*, New York, Anchor Books, 1957, p. 380–1.
21 Ibid., p. 381.
22 Ibid.
23 Verrill, *Perfume and Spices*, p. 97.
24 G. Vigarello, *Concepts of Cleanliness: Changing Attitudes in France Since the Middle Ages*, J. Birrell (trans.), Cambridge, Cambridge University Press, 1988, p. 86.
25 R. Park, *On Social Control and Collective Behaviour*, Chicago, University of Chicago Press, 1967, pp. 179–80.
26 Cited in Park, *On Social Control*, p. 180.
27 B. Connolly and R. Anderson, *First Contact: New Guinea's Highlanders Encounter the Outside World*, New York, Viking, 1987, pp. 41–2.
28 D. Howes, 'Odour in the Court', *Border/lines*, Winter 1989/90, vol. 17, pp. 28–30; E. Cohen, 'The Broken Cycle: Smell in a Bangkok Soi (Lane)', *Ethnos*, 1988, vol. 53, pp. 38–9.
29 D. Artis and S. Silvester, 'Odour Nuisance: Legal Controls', *Journal of Planning and Environment Law*, August 1986, p. 571.
30 Ibid.; J. P. Wheeler, 'Livestock Odor & Nuisance Actions vs. "Right-to-Farm" Laws: Report by Defendant Farmer's Attorney', *North Dakota Law Review*, 1992, vol. 68, no. 2, pp. 459–66.
31 P. Curran, 'Perfume Factory: Chicken Plant's Stench Drives Neighbours Crazy', *The Montreal Gazette*, Wednesday, 25 August 1993.
32 Ibid.
33 P. Black, 'No One's Sniffing at Aroma Research Now', *Businessweek*, 23 December 1991, no. 3245, pp. 82–3.
34 A. Corbin, *The Foul and the Fragrant: Odor and the French Social Imagination*, M. Kochan, R. Porter and C. Prendergast (trans.), Cambridge, Mass., Harvard University Press, 1986, p. 143.
35 R. Lifton, *The Nazi Doctors: Medical Killing and the Psychology of Genocide*, New York, Basic Books, 1986, p. 16.
36 A. Hitler, *Mein Kampf*, R. Manheim (trans.), Boston, Houghton Mifflin, 1943, p. 57.
37 Lifton, *Nazi Doctors*, p. 161.
38 Ibid., p. 31.
39 O. Lengyel, *Five Chimneys: The Story of Auschwitz*, Chicago, Ziff-Davies, 1947, p. 45, cited in H. J. Rindisbacher, *The Smell of Books: A Cultural-Historical Study of Olfactory Perception in Literature*, Ann Arbor, University of Michigan Press, 1992, p. 242.
40 Rindisbacher, *Smell of Books*, pp. 241–3.
41 J. Bezwinska (ed.), *KL Auschwitz Seen by the SS: Höss, Broad, Kremer*, New York, Howard Fertig, 1984, p. 122, cited in Rindisbacher, *Smell of Books*, p. 259.
42 Rindisbacher, *Smell of Books*, pp. 254–5. Rindisbacher provides a good discussion of the politics of smell in Nazi concentration camps, pp. 239–67.
43 Lifton, *Nazi Doctors*, p. 197.
44 Ibid., p. 196

45 L. Millu, *Smoke over Birkenau*, L. Schwartz (trans.), Philadelphia, The Jewish Publication Society, 1991, p. 157.

46 Lifton, *Nazi Doctors*, p. 345.

47 Lengyel, *Five Chimneys*, pp. 147–8, cited in Rindisbacher, *Smell of Books*, pp. 242–3.

48 B. Hyett, *In Evidence: Poems of the Liberation of Nazi Concentration Camps*, Pittsburgh, University of Pittsburgh Press, 1986, p. 8, cited in Rindisbacher, *Smell of Books*, pp. 263–4.

49 See L. Strate, 'Media and the Sense of Smell', in G. Grumpet and R. Cathcart (eds), *Inter-Media*, Oxford, Oxford University Press, 1986, pp. 428–38.

50 Mynona, 'On the Bliss of Crossing Bridges', cited in Rindisbacher, *Smell of Books*, pp. 233–4.

51 Ibid., pp. 233–5.

52 J. Gloag, *The New Pleasure*, London, George Allen & Unwin, 1933.

53 Ibid., p. 175.

54 A. Huxley, *Brave New World*, New York, Harper & Row, 1932, pp. 198–9.

55 Ibid., pp. 126, 129–30.

56 Ibid., pp. 170–1.

57 Ibid., pp. 233, 297, 302–11.

6 The aroma of the commodity

1 A. Tomlinson, 'Introduction: Consumer Culture and the Aura of the Commodity', in A. Tomlinson (ed.), *Consumption, Identity, and Style: Marketing, Meaning, and the Packaging of Pleasure*, London, Routledge, 1990, p. 9. The critique of the commodity form presented here is rooted in Marx's discussion of commodity fetishism in *Capital: A Critique of Political Economy*, vol. 1, S. Moore and E. Aveling (eds), London, Lawrence & Wishart, 1983 [1887], pp. 78–87, and Baudrillard's account of the semiology of consumption under late capitalist conditions in *For a Critique of the Political Economy of the Sign*, C. Levin (trans.), St Louis, Telos Press, 1981. See further S. Ewen, *All Consuming Images: The Politics of Style in Contemporary Culture*, New York, Basic Books, 1988.

2 R. Rosenbaum, 'Today the Strawberry, Tomorrow . . .', in N. Klein (ed.), *Culture, Curers and Contagion*, Novato, Cal., Chandler & Sharp, 1979, p. 92.

3 K. Ellison, 'Avon Ladies of the Amazon', *The Montreal Gazette*, Saturday, 2 October 1993, section B3.

4 Ibid.

5 Cited in R. Marchand, *Advertising the American Dream: Making Way for Modernity, 1920–1940*, Berkeley, University of California Press, 1985, p. 19.

6 Ibid.

7 Ibid.

8 Ibid., p. 20.

9 Ibid., p. 290.
10 Cited in Marchand, *American Dream*, p. 211.
11 See V. Vinikas, *Soft Soap, Hard Sell: American Hygiene in an Age of Advertisement*, Ames, Iowa State University Press, 1992, p. 42.
12 See C. Goodrum and H. Dalrymple, *Advertising in America: The First 200 Years*, New York, Harry N. Abrams, 1990, pp. 124, 134.
13 T. Zelman, 'Language and Perfume: A Study in Symbol-Formation', in S. R. Dana (ed.), *Advertising and Popular Culture: Studies in Variety and Versatility*, Bowling Green, Ohio, Bowling Green State University Popular Press, 1992, p. 112.
14 Ibid., p. 114.
15 On the question of this shift from reasoned argument to direct emotional appeal – or in other words, from text to image – see Tomlinson, 'Aura of the Commodity', pp. 6–11; W. Leiss, S. Kline and S. Jhaly, *Social Communication in Advertising: Persons, Products and Images of Well-Being*, Toronto, Nelson, 1988.
16 F. Naraschkewitz, 'Focus on Fragrance: Photography in Perfume Advertising', *Dragoco Report*, 1990, vol. 6, p. 222.
17 On this point see A. Synnott, 'A Sociology of Smell', *The Canadian Review of Sociology and Anthropology*, 1991, vol. 28, no. 4, pp. 449–51.
18 The following discussion on women's perfumes is based on M. Revell DeLong and E. Kersch Bye, 'Apparel for the Senses: The Use and Meaning of Fragrances', *Journal of Popular Culture*, 1990, vol. 24, no. 3, pp. 83–8; and H. Touillier-Feyrabend, 'Odeurs de séduction', *Ethnologie français*, 1989, vol. 19, no. 2, pp. 124–7.
19 Goodrum and Dalrymple, *Advertising in America*, p. 263.
20 Touillier-Feyrabend, 'Odeurs de séduction', p. 126.
21 The interests of capital in this 'expansion' of male sensuality are well brought out in W. Haug, *Critique of Commodity Aesthetics: Appearance, Sexuality and Advertising in Capitalist Society*, R. Bock (trans.), Minneapolis, University of Minnesota Press, 1986.
22 Technically, perfumes contain the highest concentration of essence, followed by eau de parfums, eau de toilettes, and finally colognes.
23 A. H. Verrill, *Perfumes and Spices*, Boston, L. C. Page, 1940, p. 96.
24 See S. Christiansen, 'The Coming Age of Aroma-Chology', *Soap, Cosmetics, Chemical Specialities*, April 1991, p. 31.
25 R. Winter, *The Smell Book*, Philadelphia, J. B. Lippincott, 1976, p. 106.
26 D. H. Powers, cited in J. S. Jellinek, *The Use of Fragrance in Consumer Products*, New York, John Wiley & Sons, 1975, p. 11.
27 Jellinek, *Fragrance in Consumer Products*, p. 10.
28 Ibid.
29 Ibid., pp. 19, 205–6.
30 Ibid., p. 199.
31 Ibid., p. 208.
32 *Maclean's*, 20 February 1989.
33 Winter, *The Smell Book*, p. 124.

34 See S. Schiffman and J. Siebert, 'New Frontiers in Fragrance Use', *Cosmetics and Toiletries*, June 1991, vol. 106, pp. 39–45.

35 'Body odor at $6,780 a gram used to collect unpaid bills', *Ottawa Citizen*, Saturday, 26 October 1991.

36 'The Nose Knows', *The Ottawa Citizen*, Thursday, 17 September 1992.

37 A. Synnott, *The Body Social*, London, Routledge, 1993, p. 203; C. Kallan, 'Scientists Say Aromas Have Major Effect on Emotions', *Los Angeles Times*, Monday, 13 May 1991, p. B3.

38 Cited in Christiansen, 'Aroma-Chology', p. 32.

39 E. Ruppel Shell, 'Chemists Whip up a Tasty Mess of Artificial Flavors', *Smithsonian*, 1986, vol. 17, no. 1, p. 79.

40 In the flavour industry flavour is defined as 'the perception of volatile flavoring substances by the nose before and during the consumption of food', hence primarily relating to the sense of smell. T. van Ejik, 'The Flavor of Cocoa and Chocolate and the Flavoring of Compound Coatings', *Dragoco Report*, 1992, no. 1, pp. 3–4.

41 Rosenbaum, 'Today the Strawberry', p. 87.

42 Ibid., p. 86.

43 Ibid., pp. 81, 83–4.

44 Ibid., p. 81.

45 Ibid., p. 87

46 R. Shell, 'Chemists', p. 80.

47 Ibid., p. 84.

48 Rosenbaum, 'Today the Strawberry', p. 88.

49 R. Shell, 'Chemists', pp. 84–6.

50 Ibid., p. 82; Rosenbaum, 'Today the Strawberry', pp. 91–2.

51 Rosenbaum, 'Today the Strawberry', p. 92.

52 Jellinek, *Fragrance in Consumer Products*, p. 133. See Rosenbaum for this situation in the flavour industry, 'Today the Strawberry', p. 82.

53 See L. Burgunder, 'Trademark Protection of Smells: Sense or Nonsense', *American Business Law Journal*, 1991, no. 29, pp. 459–80; S. King, 'Are Sounds and Scents Trade-marks in Canada?', *Business and the Law*, January 1992, pp. 6–7. For a discussion of the laws concerning the manufacture and sale of fragrances in Europe see J.-P. Pamoukdijan, *Le droit du parfum*, Paris, Librairie Générale de Droit et de Jurisprudence, 1982.

54 This is not as far-fetched as it seems. Manufacturers of vinyl shoes have been known to spray their footwear with leather scent in order to attract shoe buyers who prefer leather. Winter, *The Smell Book*, p. 107.

55 Burgunder, 'Trademark Protection of Smells', p. 468.

56 'Scenting a Generation Gap', *Harper's Magazine*, March 1992, p. 28.

57 W. S. Cain, 'Educating Your Nose', *Psychology Today*, July 1981, vol. 15, no. 7, p. 52.

58 On the condition of postmodernity see R. Kearney, *The Wake of Imagination: Toward a Postmodern Culture*, Minneapolis, University of Minnesota Press, 1988; S. Connor, *Postmodernist Culture: An*

Introduction to Theories of the Contemporary, Oxford, Blackwell, 1989; Z. Bauman, *Intimations of Postmodernity*, London, Routledge, 1991.

59 J. Baudrillard, *Simulations*, P. Foss, P. Patton and P. Beitchman (trans.), New York, Semiotext(e), 1983, p. 16.

60 Ibid., p. 3.

61 On the magic consumer kingdom of Disneyland see U. Eco, *Travels in Hyperreality*, W. Weaver (trans.), San Diego, Cal., Harcourt Brace Jovanovich, 1986, pp. 43–8.

Bibliography

Albert, J.-P., *Odeurs de sainteté: La mythologie chrétienne des aromates*, Paris, Editions de l'Ecole des Hautes Etudes en Sciences Sociales, 1990.

Almagor, U., 'The Cycle and Stagnation of Smells: Pastoralists–Fishermen Relationships in an East African Society', *RES*, vol. 14, 1987, pp. 106–21.

—— 'Odors and Private Language: Observations on the Phenomenology of Scent', *Human Studies*, vol. 13, 1990, pp. 253–74.

Amherst, A., *A History of Gardening in England*, Detroit, Singing Tree Press, 1969 [1896].

Apius and Virginia, London, Chiswick Press, 1911 [1575].

Archer-Hind, R. D., ed. and trans., *The Timaeus of Plato*, New York, Macmillan & Co., 1888.

Arnold, T., ed., *Select English Works of John Wyclif*, vol. 1, Oxford, Clarendon Press, 1869.

Artis, D. and Silvester, S., 'Odour Nuisance: Legal Controls', *Journal of Planning and Environment Law*, August 1986, pp. 565–77.

Atchley, E., *A History of the Use of Incense in Divine Worship*, London, Longmans, Green & Co., 1909.

Athenaeus, *The Deipnosophists or Banquet of the Learned*, 3 vols., C. D. Yonge, trans., London, Henry G. Bohn, 1854.

Auget, R., *Cruelty and Civilization: The Roman Games*, London, George Allen & Unwin, 1972.

Austin, T., ed., *Two Fifteenth-Century Cookery Books*, Oxford, Oxford University Press, 1964.

Ballerino Cohen, C. 'Olfactory Constitution of the Postmodern Body: Nature Challenged, Nature Adorned', in F. E. Mascia Lees and P. Sharpe, eds, *Tatoo, Torture, Mutilation and Adornment*, Binghampton, N.Y., State University of New York Press, 1992, pp. 48–78.

Balzac, H. de, *Père Goriot*, A. J. Krailsheimer, trans., Oxford, Oxford University Press, 1991 [1834].

Baudelaire, C., *The Flowers of Evil*, M. and J. Matthews, eds, NewYork, New Directions, 1989 [1857].

Baudrillard, J., *For a Critique of the Political Economy of the Sign*, C. Levin, trans., St Louis, Telos Press, 1981.

—— *Simulations*, P. Foss, P. Patton and P. Beitchman, trans., New York, Semiotext(e), 1983.

Bauman, Z., *Intimations of Postmodernity*, London, Routledge, 1991.

Bedichek, R., *The Sense of Smell*, Garden City, N.Y., Doubleday, 1960.

Beek, W. E. A. von, 'The Dirty Smith: Smell as a Social Frontier among the Kapsiki/Higi of North Cameroon and North-Eastern Nigeria', *Africa*, vol. 62, no. 1, 1992, pp. 38–58.

Bell, W., *The Great Plague in London in 1665*, London, John Lane The Bodley Head, 1924.

Black, P., 'No One's Sniffing at Aroma Research Now', *Businessweek*, no. 3245, 23 December 1991.

Bleek, W. H. I. and Lloyd, L. C., eds, *Specimens of Bushman Folklore*, London, George Allen, 1911.

Boddy, J., *Wombs and Alien Spirits: Women, Men and the Zar Cult in Northern Sudan*, Madison, University of Wisconsin Press, 1989.

Braunstein, P., 'Toward Intimacy: The Fourteenth and Fifteenth Centuries', in G. Duby, ed., *A History of Private Life, II: Revelations of the Medieval World*, A. Goldhammer, trans., Cambridge, Mass., Belknap Press, 1988, pp. 535–630.

Brill, A. A., 'The Sense of Smell in the Neuroses and Psychoses', *The Psychoanalytic Quarterly*, vol. 1, 1932, pp. 7–42.

Buckley, T. and Gottlieb, A., eds, *Blood Magic: The Anthropology of Menstruation*, Berkeley, University of California Press, 1988.

Burgunder, L., 'Trademark Protection of Smells: Sense or Nonsense', *American Business Law Journal*, vol. 29, 1991, pp. 459–80.

Burton, E., *The Elizabethans at Home*, London, Secker & Warburg, 1958.

Cain, W. S., 'Educating Your Nose', *Psychology Today*, vol. 15, no. 7, July 1981, pp. 48–56.

Calame-Griaule, G., *Words and the Dogon World*, D. Lapin, trans., Philadelphia, Institute for the Study of Human Issues, 1986.

Campbell, A., *The Black Death and Men of Learning*, New York, AMC Press, 1966.

Campbell, J. D., ed., *The Poetical Works of Samuel Taylor Coleridge*, London, MacMillan, 1924.

Carterette, E. C. and Friedman, M. P., eds, *Handbook of Perception*, Vol. 6A: *Tasting and Smelling*, New York, Academic Press, 1978.

Cashmore, E. E., *The Logic of Racism*, London, Allen & Unwin, 1987.

Chaucer, G., *The Canterbury Tales*, D. Wright, trans., Oxford, Oxford University Press, 1985.

Christiansen, S., 'The Coming Age of Aroma-Chology', *Soap, Cosmetics, Chemical Specialties*, April 1991, pp. 30–2.

Classen, C. *Inca Cosmology and the Human Body*, Salt Lake City, University of Utah Press, 1993.

—— *Worlds of Sense: Exploring the Senses in History and Across Cultures*, London, Routledge, 1993.

—— and D. Howes, 'Aromatherapy in the Andes', *Dragoco Report*, vol. 40, no. 6, 1993, pp. 213–27.

Cobham Brewer, E., *A Dictionary of Miracles*, Detroit, Gale Research Company, 1966.

Cohen, E., 'The Broken Cycle: Smell in a Bangkok Soi (Lane)', *Ethnos*, vol. 53, 1988, pp. 37–49.

Condillac, E. B. de, *Treatise on the Sensations*, G. Carr, trans., Los Angeles, University of Southern California Press, 1930.

Connolly, B. and Anderson, R., *First Contact: New Guinea's Highlanders Encounter the Outside World*, New York, Viking, 1987.

Connor, S., *Postmodernist Culture: An Introduction to Theories of the Contemporary*, Oxford, Blackwell, 1989.

Corbin, A., *The Foul and the Fragrant: Odor and the French Social Imagination*, M. Kochan, R. Porter and C. Prendergast, trans., Cambridge, Mass., Harvard University Press, 1986.

—— 'Histoire et anthropologie sensorielle', *Anthropologie et Sociétés*, vol. 14, no. 2, 1990, pp. 13–24.

Corcopino, J., *Daily Life in Ancient Rome*, E. Lorimer, trans., New Haven, Conn., Yale University Press, 1940.

Crocker, J. C., *Vital Souls: Bororo Cosmology, Natural Symbolism, and Shamanism*, Tucson, University of Arizona Press, 1985.

Curran, P., 'Perfume Factory: Chicken Plant's Stench Drives Neighbours Crazy', *The Montreal Gazette*, Wednesday, 25 August 1993.

Darwin, C., *The Descent of Man and Selection in Relation to Sex*, New York, D. Appleton, 1898.

Defoe, D., *A Journal of the Plague Year*, London, J. M. Dent, 1928 [1722].

Detienne, M., *The Gardens of Adonis: Spices in Greek Mythology*, J. Lloyd, trans., Atlantic Highlands, N.J., The Humanities Press, 1977.

Dollard, J., *Caste and Class in a Southern Town*, New York, Anchor Books, 1957.

Doty, R., 'The Role of Olfaction in Man: Sense or Nonsense?', in S. H. Bartley, ed., *Perception in Everyday Life*, New York, Harper & Row, 1972, pp. 143–57.

—— 'Olfactory Communication in Humans', *Chemical Senses*, vol. 6, no. 4, 1981 pp. 351–76.

DuBois, P., *Centaurs and Amazons: Women and the Pre-History of the Great Chain of Being*, Ann Arbor, University of Michigan Press, 1982.

Duckworth, G., ed., *The Complete Roman Drama*, 2 vols., New York, Random House, 1942.

Dupire, M., 'Nomination, réincarnation et/ou ancêtre tutélaire? Un mode de survie: L'exemple des serer ndout (Senegal)', *L'Homme*, vol. 22, no. 1, 1982, pp. 5–31.

—— 'Des goûts et des odeurs: classifications et universaux', *L'Homme*, vol. 27, no. 4, 1987, pp. 5–24.

Eastwood, B. S., 'Galen on the Elements of Olfactory Sensation', *Rheinisches Museum für Philologie*, vol. 124, 1981, pp. 268–89.

Eco, U., *Travels in Hyperreality*, W. Weaver, trans., San Diego, Cal., Harcourt Brace Jovanovich, 1986.

Ejik, T. van, 'The Flavor of Cocoa and Chocolate and the Flavoring of Compound Coatings', *Dragoco Report*, no. 1, 1992, pp. 3–25.

Elias, N., *The Civilizing Process*, vol. 1: *The History of Manners*, E. Jephcott, trans., New York, Pantheon, 1982.

Ellis, H., *Studies in the Psychology of Sex*, vol. 1, part 3, New York, Random House, 1942 [1899].

Ellison, K., 'Avon Ladies of the Amazon', *The Montreal Gazette*, Saturday, 2 October 1993.

Endicott, K., *Batek Negrito Religion: The World-View and Rituals of a Hunting and Gathering People of Peninsular Malaysia*, Oxford, Clarendon Press, 1979.

Engen, T., *Odor Sensation and Memory*, New York, Praeger, 1991.

Ewen, S., *All Consuming Images: The Politics of Style in Contemporary Culture*, New York, Basic Books, 1988.

Fauré, P., *Parfums et aromates de l'antiquité*, Paris, Fayard, 1987.

Forbes, H., *A Naturalist's Wanderings in the Eastern Archipelago*, London, Sampson Low, Marston, Searle & Rivington, 1885.

Freud, S., *Civilization and its Discontents*, J. Strachey, trans., New York, W. W. Norton, 1961.

Friedländer, L., *Roman Life and Manners Under the Early Empire*, J. Freese and L. Magnus, trans., London, Routledge & Kegan Paul, 1908.

Gavin, H., *Sanitary Ramblings*, London, John Churchill, 1848.

Gebhart-Sayer, A., 'The Geometric Designs of the Shipibo-Conibo in Ritual Context', *Journal of Latin American Lore*, vol. 11, no. 2, 1985, pp. 143–75.

Gell, A., 'Magic, Perfume, Dream . . .', in I. M. Lewis, ed., *Symbols and Sentiments*, London, Academic Press, 1977, pp. 25–38.

Genders, R., *A History of Scent*, London, Hamish Hamilton, 1972.

Gloag, J., *The New Pleasure*, London, George Allen & Unwin, 1933.

Goodrum, C. and Dalrymple, H., *Advertising in America: The First 200 Years*, New York, Harry N. Abrams, 1990.

Grant, M., *The Fall of the Roman Empire: A Reappraisal*, Randar, Penn., Annenberg School Press, 1976.

Greek Anthology, The, vol. 1, W. Paton, trans., London, William Heinemann, 1960.

Grene, D. and Lattimore, R., eds, *Sophocles*, Chicago, University of Chicago Press, 1954.

Griffiths, J. G., ed., *Plutarch's 'De Iside et Osiride'*, Cardiff, University of Wales Press, 1970.

Groom, N., *Frankincense and Myrrh: A Study of the Arabian Incense Trade*, London, Longman, 1981.

Guérin, P., *Vies des saints*, 7th edn, vol. 5, Paris, Bloud & Barral, 1878.

Hall, E. T., *The Hidden Dimension*, Garden City, N.Y., Anchor, 1969.

Hamilton, D., 'Chirac's Remark on Immigrants Touches Raw Nerve in French Politics', *The Montreal Gazette*, Saturday, 22 June 1991.

Harper, R., Bate Smith, E. C. and Land, D. G., *Odour Description and Odour Classification*, London, J. & A. Churchill, 1968.

Harrington, J., *A New Discourse of a Stale Subject called the Metamorphosis of Ajax*, E. Donno, ed., London, Routledge & Kegan Paul, 1962 [1596].

Harvey, E. R., *The Inward Wits: Psychological Theory in the Middle Ages and the Renaissance*, London, The Warburg Institute, 1975.

Harvey, S., 'The Fragrance of Sanctity: Incense and Spirituality in the Early Byzantine East', Dumbarton Oaks Public Lecture, 11 March 1992.

Haug, W., *Critique of Commodity Aesthetics: Appearance, Sexuality and Advertising in Capitalist Society*, R. Bock, trans., Minneapolis, University of Minnesota Press, 1986.

Henderson, J. B., *The Development and Decline of Chinese Cosmology*, New York, Columbia University Press, 1984.

Hitler, A., *Mein Kampf*, R. Manheim, trans., Boston, Houghton Mifflin, 1943.

Hogg, J., *London as It Is*, London, John Macrone, 1837.

Homer, *The Iliad*, 2 vols., A. Murray, trans., London, William Heinemann, 1924.

—— *The Homeric Hymns*, D. Hines, trans., New York, Atheneum, 1972.

Horace, *Works of Horace*, C. Smart, trans., London, George Bell & Sons, 1885.

Howell, S., *Society and Cosmos: Chewong of Peninsular Malaysia*, Oxford, Oxford University Press, 1984.

Howes, D., 'On the Odour of the Soul: Spatial Representation and Olfactory Classification in Eastern Indonesia and Western Melanesia', *Bijdragen tot de Taal-, Land- en Volkenkunde*, vol. 144, 1988, pp. 84–113.

—— 'Odour in the Court', *Border/lines*, vol. 17, Winter 1989/90, pp. 28–30.

—— 'Olfaction and Transition', in D. Howes, ed., *The Varieties of Sensory Experience: A Sourcebook in the Anthropology of the Senses*, Toronto, University of Toronto Press, 1991, pp. 128–47.

—— 'Sense and Non-Sense in Contemporary Ethno/Graphic Practice and Theory', *Culture*, vol. 11, nos. 1–2, 1991, pp. 65–76.

—— ed., *The Varieties of Sensory Experience: A Sourcebook in the Anthropology of the Senses*, Toronto, University of Toronto Press, 1991.

—— and Lalonde, M., 'The History of Sensibilities: Of the Standard of Taste in Mid-Eighteenth Century England and the Circulation of Smells in Post-Revolutionary France', *Dialectical Anthropology*, vol. 16, 1992, pp. 125–35.

Hughes-Halet, L., *Cleopatra: Histories, Dreams and Distortions*, New York, Harper & Row, 1990.

Hugo, V., *Les Misérables*, C. Wilbour, trans., New York, Modern Library, n.d.

Hutchinson, F. E., ed., *The Works of George Herbert*, Oxford, Clarendon Press, 1941.

Huxley, A., *Brave New World*, New York, Harper & Row, 1932.

Huysmans, J.-K., *Against Nature*, R. Baldwick, trans., London, Penguin, 1959 [1884].

Ingram, W. G. and Redpath, T., eds, *Shakespeare's Sonnets*, London, University of London Press, 1964.

James, W., *The Listening Ebony: Moral Knowledge, Religion and Power Among the Uduk of Sudan*, Oxford, Clarendon Press, 1988.

Jellinek, J. S., *The Use of Fragrance in Consumer Products*, New York, John Wiley & Sons, 1975.

Juvenal, 'Satires', in *Juvenal and Persius*, G. Ramsay, trans., London, Heinemann, 1957.

Kahn, C. H., *The Art and Thought of Heraclitus: An Edition of the Fragments with Translation and Commentary*, Cambridge, Cambridge University Press, 1979.

Kahn, M., *Always Hungry, Never Greedy: Food and the Expression of Gender in a Melanesian Society*, Cambridge, Cambridge University Press, 1986.

Kallan, C., 'Scientists Say Aromas Have Major Effect on Emotions', *Los Angeles Times*, 13 May, 1991.

Kanafani, A., *Aesthetics and Ritual in the United Arab Emirates: The Anthropology of Food and Personal Adornment among Arabian Women*, Beirut, American University of Beirut, 1983.

Kant, I., *Anthropology from a Pragmatic Point of View*, V. L. Dowdell, trans., Carbondale and Edwardsville, Southern Illinois University Press, 1978 [1798].

Kearney, R. *The Wake of Imagination: Toward a Postmodern Culture*, Minneapolis, University of Minnesota Press, 1988.

Kiernan, M., ed., *Sir Francis Bacon: The Essayes or Counsels, Civill and Morall*, Oxford, Clarendon Press, 1985 [1625].

King, S., 'Are Sounds and Scents Trademarks in Canada?', *Business & The Law*, January 1992, pp. 6–7.

Küchler, S., 'The Epidemiology of Imagery and the Epistemology of Smell in Northern New Ireland', paper presented at 87th Annual Meeting of the American Anthropological Association, 1988.

Lamb, F. B., *Wizard of the Upper Amazon*, Boston, Hougton Mifflin, 1975.

Lambek, M., *Human Spirits: A Cultural Account of Trance in Mayotte*, Cambridge, Cambridge University Press, 1981.

Largey, G. P. and Watson, D. R., 'The Sociology of Odors', *The American Journal of Sociology*, vol. 77, no. 6, 1977, pp. 1021–34.

Le Guérer, A., *Scent: The Mysterious and Essential Powers of Smell*, R. Miller, trans., New York, Turtle Bay Books, 1992.

Leacock, S. and Leacock, R., *Spirits of the Deep: A Study of an Afro-Brazilian Cult*, Garden City, N.Y., Doubleday Natural History Press, 1972.

Leenhardt, M., *Do Kamo: Person and Myth in the Melanesian World*, B. Miller Gulati, trans., Chicago, University of Chicago Press, 1979.

Leiss, W., Kline, S. and Jhaly, S., *Social Communication in Advertising: Persons, Products and Images of Well-Being*, Toronto, Nelson, 1988.

Lévi-Strauss, C., *The Raw and the Cooked: Introduction to a Science of Mythology*, vol. 1, J. and D. Weightman, trans., New York, Harper & Row, 1969.

Lewis, G., *Knowledge of Illness in a Sepik Society: A Study of the Gnau, New Guinea*, London, Athlone Press, 1975.

Lifton, R., *The Nazi Doctors: Medical Killing and the Psychology of Genocide*, New York, Basic Books, 1986.

Lilja, S., *The Treatment of Odours in the Poetry of Antiquity*, Commentationes Humanarum Litterarum, 49, Helsinki, Societas Scientiarum Fennica, 1972.

Lindsay, A. D., ed., *Socratic Discourses by Plato and Xenophon*, London, J. M. Dent, 1910.

Lives of the Later Caesars: The First Part of the Augustan History, A. Birley, trans., Harmondsworth, Penguin, 1976.

Lucan, *The Civil War*, vol. 1, J. Duff, trans., London, William Heinemann, 1957.

Lucretius, *On the Nature of Things*, C. Bailey, trans., Oxford, Clarendon Press, 1910.

McLaughlin, T., *Coprophilia or A Peck of Dirt*, London, Cassell, 1971.

Majno, G., *The Healing Hand: Man and Wound in the Ancient World*, Cambridge, Mass., Harvard University Press, 1975.

Malinowski, B., *The Sexual Life of Savages in North-Western Melanesia*, London, Harcourt, Brace & World, 1929.

—— *Coral Gardens and Their Magic*, vol. 1, Bloomington, Indiana University Press, 1965.

Marchand, R., *Advertising the American Dream: Making Way for Modernity, 1920–1940*, Berkeley, University of California Press, 1985.

Martial, *Epigrams*, W. Kerr, trans., Cambridge, Mass., Harvard University Press, 1961.

Martin, L. C., ed., *The Poetical Works of Robert Herrick*, Oxford, Clarendon Press, 1956.

Marx, K., *Capital: A Critique of Political Economy*, vol. 1, S. Moore and E. Aveling, eds, London, Lawrence & Wishart, 1983 [1887].

Maugham, W. S., *On A Chinese Screen*, Oxford, Oxford University Press, 1927.

Meyer, J. J., *Sexual Life in Ancient India*, Delhi, Banarsidass, 1971.

Miller, J. I., *The Spice Trade of the Roman Empire*, Oxford, Clarendon Press, 1969.

Millu, L., *Smoke Over Birkenau*, L. Schwartz, trans., Philadelphia, The Jewish Publication Society, 1991.

Muir, F., *An Irreverent and Almost Complete Social History of the Bathroom*, New York, Stein & Day, 1983.

Nakbi, J.-L., 'Sémiologie des odeurs rituelles: Rites et rythmes olfactifs dans le sud tunisien', *Cahiers de Sociologie Economique et Culturelle*, vol. 4, 1985, pp. 117–36.

Naraschkewitz, F., 'Focus on Fragrance: Photography in Perfume Advertising', *Dragoco Report*, no. 6, 1990, pp. 222–33.

Neame, A., *The Holy Maid of Kent: The Life of Elizabeth Barton*, London, Hodder & Stoughton, 1971.

Needham, J., *Science and Civilisation in China*, vol. 2, Cambridge, Cambridge University Press, 1969.

Nichols, J., *The Progresses of Queen Elizabeth*, London, John Nichols & Son, 1823.

Nielsen, K., *Incense in Ancient Israel*, Leiden, E. J. Brill, 1986.

Nordau, M., *Degeneration*, New York, D. Appleton, 1902.

Norton, T., *Ordinal of Alchemy*, J. Reidy, ed., Oxford, Oxford University Press, 1975.

Oates, W. and O'Neill, E., Jr., eds, *The Complete Greek Drama*, 2 vols., New York, Random House, 1938.

Orwell, G., *The Road To Wigan Pier*, London, Victor Gollancz, 1937.

Ovid, *Metamorphoses*, vol. 1, F. Miller, trans., London, William Heinemann, 1960.

Palgrave, W. G., *Narrative of a Year's Journey Through Central and Eastern Arabia*, vol. 2, London, Macmillan & Company, 1866.

Palmer, R., 'In Bad Odour: Smell and its Significance in Medicine from Antiquity to the Seventeenth Century', in R. Porter and W. F. Bynum, eds, *Medicine and the Five Senses*, Cambridge, Cambridge University Press, 1993, pp. 61–8.

Pamoukdijan, J., *Le droit du parfum*, Paris, Librairie Générale de Droit et de Jurisprudence, 1982.

Pandya, V., 'Above the Forest: A Study of Andamanese Ethnoamenology, Cosmology and the Power of Ritual', Ph.D. thesis, University of Chicago, 1987.

—— 'Movement and Space: Andamanese Cartography', *American Ethnologist*, vol. 17, no. 4, 1991, pp. 775–97.

Panoff, M., 'The Notion of Time Among the Maenge People of New Britain', *Ethnology*, vol. 8, 1969, pp. 153–66.

Parfitt, G., ed., *The Complete Poems of Ben Johnson*, Harmondsworth, Penguin, 1975.

Park, R., *On Social Control and Collective Behaviour*, Chicago, University of Chicago Press, 1967.

Petit-Skinner, S., *The Nauruans*, San Francisco, Macduff Press, 1981.

Petronius, 'Satyricon', in *Petronius*, M. Heseltine, trans., London, William Heinemann, 1919.

Peyrot, M., 'La plainte du MRAP contre M. Chirac: Le procès des "odeurs" ', *Le Monde*, 31 January, 1992.

Piesse, C. H., ed., *Piesse's Art of Perfumery*, London, Piesse & Ludin, 1891.

Plato, 'Symposium', in *Plato*, vol. 5, W. Lamb, trans., London, William Heinemann, 1925.

Plautus, 'Poenulus', in *Plautus*, vol. 4, P. Nixon, trans., London, William Heinemann, 1959.

Pliny, *Natural History*, H. Rackham, trans., London, William Heinemann, 1960.

Plutarch, *The Age of Alexander: Nine Greek Lives*, I. Scott-Kilvert, trans., Harmondsworth, Penguin, 1973.

—— *Makers of Rome: Nine Lives*, I. Scott-Kilvert, trans., Harmondsworth, Penguin, 1965.

—— *Fall of the Roman Republic*, R. Warner, trans., Harmondsworth, Penguin, 1972.

Porkert, M., *The Theoretical Foundations of Chinese Medicine*, Cambridge, Mass., MIT Press, 1974.

Porteous, D., *Landscapes of the Mind: Worlds of Sense and Metaphor*, Toronto, University of Toronto Press, 1990.

Poupon, P., *Mes dégustations littéraires: L'odorat et le goût chez les écrivains*, Dijon-Quetigny, Imprimerie Darantière, 1979.

Proust, M., *Remembrance of Things Past*, Vol. I: *Swann's Way*, C. K. Moncrieff and T. Kilmartin, trans., New York, Random House, 1981 [1913].

Pullar, P., *Consuming Passions: A History of English Food and Appetite*, London, Hamish Hamilton, 1970.

Radcliffe-Brown, A. R., *The Andaman Islanders*, New York, Free Press, 1964.

Reichel-Dolmatoff, G., *Amazonian Cosmos: The Sexual and Religious Symbolism of the Tukano Indians*, Chicago, University of Chicago Press, 1971.

—— *Beyond the Milky Way: Hallucinatory Images of the Tukano Indians*, Los Angeles, UCLA Latin American Center Publications, 1978.

—— 'Desana Animal Categories, Food Restrictions, and the Concept of Color Energies', *Journal of Latin American Lore*, vol. 4, no. 2, 1978, pp. 243–91.

—— 'Brain and Mind in Desana Shamanism', *Journal of Latin American Lore*, vol. 7, no. 1, 1981, pp. 73–98.

—— *Basketry as Metaphor: Arts and Crafts of the Desana Indians of the Northwest Amazon*, Los Angeles, Occasional Papers of the Museum of Cultural History, University of California, 1985.

—— 'Tapir Avoidance in the Colombian Northwest Amazon', in G. Urton, ed., *Animal Myths and Metaphors in South America*, Salt Lake City, University of Utah Press, 1985, pp. 107–43.

Revell DeLong, M. and Kersch Bye, E., 'Apparel for the Senses: The Use and Meaning of Fragrances', *Journal of Popular Culture*, vol. 24, no. 3, 1990, pp. 81–8.

Reynolds, R., *Cleanliness and Godliness*, London, George Allen & Unwin, 1943.

Rindisbacher, H. J., *The Smell of Books: A Cultural-Historical Study of Olfactory Perception in Literature*, Ann Arbor, University of Michigan Press, 1992.

Roheim, G., *Psychoanalysis and Anthropology*, New York, International Universities Press, 1950.

Roseman, M., *Healing Sounds from the Malaysian Rainforest: Temiar Music and Medicine*, Berkeley, University of California Press, 1991.

Ross, W., ed., *The Works of Aristotle*, vol. VII: *Problemata*, Oxford, Clarendon Press, 1927.

Rowse, A. L., ed., *The Annotated Shakespeare*, New York, Clarkson N. Potter, 1978.

Ruppel Shell, E., 'Chemists Whip up a Tasty Mess of Artificial Flavors', *The Smithsonian*, vol. 17, no. 1, 1986, pp. 79–88.

Sacks, O., *The Man Who Mistook His Wife for a Hat*, London, Duckworth, 1987.

Ste Croix, G. de, *The Class Struggle in the Ancient World*, Ithaca, N.Y., Cornell University Press, 1981.

Schaal, B. and Porter, R., ' "Microsmatic Humans" Revisited: The Generation and Perception of Chemical Signals', *Advances in the Study of Behavior*, vol. 20, 1991, pp. 135–9.

Schiffman, S. and Siebert, J, 'New Frontiers in Fragrance Use', *Cosmetics and Toiletries*, no. 106, June 1991, pp. 39–45.

Schivelbusch, W., *Tastes of Paradise: A Social History of Spices, Stimulants, and Intoxicants*, D. Jacobson, trans., New York, Pantheon Books, 1992.

Seeger, A., *Nature and Society in Central Brazil: The Suya Indians of Mato Grosso*, Cambridge, Mass., Harvard University Press, 1981.

—— 'Anthropology and Odor: From Manhattan to Mato Grosso', *Perfumer & Flavorist*, vol. 13, 1988, pp. 41–8.

Shawcross, J. T., ed., *The Complete Poetry of John Donne*, New York, New York University Press, 1968.

Shulman, D.,'The Scent of Memory in Hindu South India', *RES*, vol. 13, 1987, pp. 123–33.

Siegel, J. T., 'Images and Odors in Javanese Practices Surrounding Death', *Indonesia*, vol. 36, no. 1, 1983, pp. 1–15.

Slater, W., ed., *Dining in a Classical Context*, Ann Arbor, University of Michigan Press, 1991.

Smith, W., *Dictionary of Greek and Roman Antiquities*, London, Walton & Maberly, 1853.

Spenser, E., *The Yale Edition of the Shorter Poems of Edmund Spenser*, W. Oram, E. Bjorvand, R. Bond, T. Cain, A. Dunlop and R. Schell, eds, New Haven, Conn., Yale University Press, 1989.

Stoddart, D. M., *The Scented Ape*, Cambridge, Cambridge University Press, 1990.

Stoller, P. and Olkes, C., *In Sorcery's Shadow: A Memoir of Apprenticeship Among the Songhay of Niger*, Chicago, University of Chicago Press, 1987.

Strate, L., 'Media and the Sense of Smell', in G. Grumpet and R. Cathcart, eds, *Inter-Media*, Oxford, Oxford University Press, 1986.

Suetonius, *The Twelve Caesars*, R. Graves, trans., Harmondsworth, Penguin, 1976.

Swantz, M.-L., *Ritual and Symbol in Transitional Zaramo Society with Special Reference to Women*, Uppsala, Almquist & Wiksells, 1970.

Swift, J., *Directions to Servants and Miscellaneous Pieces*, H. Davis, ed., Oxford, Blackwell, 1959 [1745].

—— *Gulliver's Travels*, R. Greenberg, ed., New York, W. W. Norton, 1961 [1726].

Synnott, A., 'Puzzling Over the Senses: From Plato to Marx', in D. Howes, ed., *The Varieties of Sensory Experience: A Sourcebook in the Anthropology of the Senses*, Toronto, University of Toronto Press, 1991, pp. 61–76.

—— 'A Sociology of Smell', *The Canadian Review of Sociology and Anthropology*, vol. 28, no. 4, 1991, pp. 437–59.

—— *The Body Social: Symbolism, Self and Society*, London, Routledge, 1993.

—— 'Roses, Coffee, and Lovers: The Meanings of Smell', unpublished manuscript, 1993.

Theophrastus, *Enquiry into Plants and Minor Works on Odours and Weather Signs*, vol. 2, A. Hort, trans., London, William Heinemann, 1961.

Thompson, C. J. S., *The Mystery and Lure of Perfume*, London, John Lane The Bodley Head, 1927.

Thorndike, L., *A History of Magic and Experimental Science*, vol. 7, New York, Colombia University Press, 1958.

Toller, S. van and Dodd, G., eds, *Perfumery: The Psychology and Biology of Fragrance*, London, Chapman & Hall, 1988.

Tomlinson, A., 'Introduction: Consumer Culture and the Aura of the Commodity', in A. Tomlinson, ed., *Consumption, Identity, and Style: Marketing, Meaning, and the Packaging of Pleasure*, London, Routledge, 1990.

Touillier-Feyrabend, H., 'Odeurs de séduction', *Ethnologie française*, vol. 19, no. 2, 1989, pp. 123–9.

Toynbee, J. M. C., *Death and Burial in the Roman World*, Ithaca, N.Y., Cornell University Press, 1971.

Verrill, A. H., *Perfumes and Spices*, Boston, L. C. Page, 1940.

Vigarello, G., *Concepts of Cleanliness: Changing Attitudes in France since the Middle Ages*, J. Birrell, trans., Cambridge, Cambridge University Press, 1988.

Vinikas, V., *Soft Soap, Hard Sell: American Hygiene in an Age of Advertisement*, Ames, Iowa State University Press, 1992.

Virgil, *The Aeneid*, J. Mantinband, trans., New York, Frederick Ungar, 1964.

Vogt, E. Z., *Tortillas for the Gods: A Symbolic Analysis of Zinacanteco Rituals*, Cambridge, Mass., Harvard University Press, 1976.

Wall, L. L., *Hausa Medicine*, Durham, N.C., Duke University Press, 1988.

Westermarck, E. A., *Ritual and Belief in Morocco*, vol. 1, London, Macmillan, 1926.

Wheeler, J. P., 'Livestock Odor & Nuisance Actions vs. "Right-to-Farm" Laws: Report by Defendant Farmer's Attorney', *North Dakota Law Review*, vol. 68, no. 2, 1992, pp. 459–66.

White, E. B., *The Fox of Peapack and Other Poems*, New York, Harper & Brothers, 1938.

Wilbert, W. 'The Pneumatic Theory of Female Warao Herbalists', *Social Sciences and Medicine*, vol. 25, no. 10, 1987, pp. 1139–46.

Wilde, O., *The Picture of Dorian Gray*, New York, Modern Library, 1985 [1891].

Williams, H., ed., *The Poems of Jonathan Swift*, Oxford, Clarendon Press, 1937.

Wilson, F. P., ed., *The Plague Pamphlets of Thomas Dekker*, Oxford, Clarendon Press, 1925.

—— *The Plague in Shakespeare's London*, Oxford, Oxford University Press, 1927.

Winter, R., *The Smell Book*, Philadelphia, J. B. Lippincott, 1976.

Wright, L., *Clean and Decent: The Fascinating History of the Bathroom and the Water Closet*, Toronto, University of Toronto Press, 1960.

Zelman, T., 'Language and Perfume: A Study in Symbol-Formation', in S. R. Dana, ed., *Advertising and Popular Culture: Studies in Variety and Versatility*, Bowling Green, Ohio, Bowling Green State University Popular Press, 1992, pp. 109–14.

Zola, E., *Nana*, New York, Collier, 1962 [1880].

Index